Undergraduate Texts in Mathematics

Springer
New York
Berlin
Heidelberg
Barcelona
Budapest
Hong Kong
London
Milan
Paris
Santa Clara
Singapore
Tokyo

Undergraduate Texts in Mathematics

Anglin: Mathematics: A Concise History and Philosophy.
Readings in Mathematics.
Anglin/Lambek: The Heritage of Thales.
Readings in Mathematics.
Apostol: Introduction to Analytic Number Theory. Second edition.
Armstrong: Basic Topology.
Armstrong: Groups and Symmetry.
Axler: Linear Algebra Done Right.
Bak/Newman: Complex Analysis. Second edition.
Banchoff/Wermer: Linear Algebra Through Geometry. Second edition.
Berberian: A First Course in Real Analysis.
Brémaud: An Introduction to Probabilistic Modeling.
Bressoud: Factorization and Primality Testing.
Bressoud: Second Year Calculus.
Readings in Mathematics.
Brickman: Mathematical Introduction to Linear Programming and Game Theory.
Browder: Mathematical Analysis: An Introduction.
Cederberg: A Course in Modern Geometries.
Childs: A Concrete Introduction to Higher Algebra. Second edition.
Chung: Elementary Probability Theory with Stochastic Processes. Third edition.
Cox/Little/O'Shea: Ideals, Varieties, and Algorithms. Second edition.
Croom: Basic Concepts of Algebraic Topology.
Curtis: Linear Algebra: An Introductory Approach. Fourth edition.
Devlin: The Joy of Sets: Fundamentals of Contemporary Set Theory. Second edition.
Dixmier: General Topology.
Driver: Why Math?
Ebbinghaus/Flum/Thomas: Mathematical Logic. Second edition.
Edgar: Measure, Topology, and Fractal Geometry.Elaydi: Introduction to Difference Equations.
Exner: An Accompaniment to Higher Mathematics.
Fischer: Intermediate Real Analysis.
Flanigan/Kazdan: Calculus Two: Linear and Nonlinear Functions. Second edition.
Fleming: Functions of Several Variables. Second edition.
Foulds: Combinatorial Optimization for Undergraduates.
Foulds: Optimization Techniques: An Introduction.
Franklin: Methods of Mathematical Economics.
Hairer/Wanner: Analysis by Its History.
Readings in Mathematics.
Halmos: Finite-Dimensional Vector Spaces. Second edition.
Halmos: Naive Set Theory.
Hämmerlin/Hoffmann: Numerical Mathematics.
Readings in Mathematics.
Hilton/Holton/Pedersen: Mathematical Reflections: In a Room with Many Mirrors.
Iooss/Joseph: Elementary Stability and Bifurcation Theory. Second edition.
Isaac: The Pleasures of Probability.
Readings in Mathematics.
James: Topological and Uniform Spaces.
Jänich: Linear Algebra.
Jänich: Topology.
Kemeny/Snell: Finite Markov Chains.
Kinsey: Topology of Surfaces.
Klambauer: Aspects of Calculus.
Lang: A First Course in Calculus. Fifth edition.
Lang: Calculus of Several Variables. Third edition.
Lang: Introduction to Linear Algebra. Second edition.
Lang: Linear Algebra. Third edition.
Lang: Undergraduate Algebra. Second edition.
Lang: Undergraduate Analysis.
Lax/Burstein/Lax: Calculus with Applications and Computing. Volume 1.
LeCuyer: College Mathematics with APL.
Lidl/Pilz: Applied Abstract Algebra.
Macki-Strauss: Introduction to Optimal Control Theory.
Malitz: Introduction to Mathematical Logic.

(continued after index)

B.A. Sethuraman

Rings, Fields, and Vector Spaces

An Introduction to Abstract Algebra via Geometric Constructibility

B.A. Sethuraman
Department of Mathematics
California State University Northridge
Northridge, CA 91330
USA

Mathematics Subject Classification (1991): 12-01, 13-01

Library of Congress Cataloging-in-Publication Data
Sethuraman, B.A.
Rings, fields, and vector spaces : an introduction to abstract algebra via geometric constructibility / B.A. Sethuraman.
p. cm. — (Undergraduate texts in mathematics)
Includes bibliographical references (p. 185–186) and index.
ISBN 0-387-94848-1 (alk. paper)
1. Algebra, Abstract. I. Title. II. Series.
QA162.S44 1996
512′.02—DC20 96-32220

Printed on acid-free paper.

Producion managed by Lesley Poliner; manufacturing supervised by Jacqui Ashri.
Camera-ready copy prepared by the author.
Printed and bound by Maple-Vail Book Manufacturing Group, York, PA.
Printed in the United States of America.

9 8 7 6 5 4 3 2 1

ISBN 0-387-94848-1 Springer-Verlag New York Berlin Heidelberg SPIN 10545117

This book is dedicated to Prabha,
who gave me so much,
and taught me so much more.

Preface

This book is an attempt to communicate to undergraduate mathematics majors my enjoyment of abstract algebra. It grew out of a course offered at California State University, Northridge, in our teacher preparation program, titled Foundations of Algebra, that was intended to provide an advanced perspective on high-school mathematics. When I first prepared to teach this course, I needed to select a set of topics to cover. The material that I selected would clearly have to have some bearing on school-level mathematics, but at the same time would have to be substantial enough for a university-level course. It would have to be something that would give the students a perspective into abstract mathematics, a feel for the conceptual elegance and grand simplifications brought about by the study of structure. It would have to be of a kind that would enable the students to develop their creative powers and their reasoning abilities. And of course, it would all have to fit into a sixteen-week semester.

The choice to me was clear: we should study constructibility. The mathematics that leads to the proof of the nontrisectibility of an arbitrary angle is beautiful, it is accessible, and it is worthwhile. Every teacher of mathematics would profit from knowing it.

Now that I had decided on the topic, I had to decide on how to develop it. All the students in my course had taken an earlier course on sets and functions, but many had not progressed any further into abstract mathematics. What I needed to do, therefore, was to develop enough algebra to lead to the proofs of the nonconstructibility results without getting bogged down in technicalities. But since this course was going to be the only algebra course that several of my students would take, the material I developed needed be rich enough so that everybody would get a good sense of what the subject was all about.

Given this goal for the course, I set out to find a textbook. There certainly is a wealth of rather excellent textbooks on introductory abstract algebra, but they seem to be designed with a different purpose in mind: to develop technical mastery of the subject. As such, they delve into the details of the subject, rather than focusing on an overview. For me to have culled from existing textbooks the mathematics that I wanted to cover in my course at the level that I wanted to cover it would have been a horrendous task. I decided instead to write my own book.

This book has been written in a conversational style, a style that mirrors my own approach to teaching. The focus is on exposition, on conveying mathematical intuition to an audience that will have careers in mathematics, but for the most part will not go on to get a Ph.D. in mathematics. Familiarity with the material is developed by exposing the students to lots of examples; sacrificing, if necessary, the desire to prove lots of theorems. The text is peppered liberally with questions, designed to encourage the students to learn the subject by thinking through the material themselves. This is particularly true of the sections that deal with examples: many of the questions asked within these examples could serve just as well as formal exercises.

The book begins with an essay on how to learn mathematics, a topic that I feel is well worth spending some time on in introductory courses in abstract mathematics. This is followed in Chapter 1 by a study of divisibility in the integers, and in Chapter 2 by a general introduction to rings and fields. Vector spaces are introduced in Chapter 3 so as to make it possible to measure degrees of field extensions. Chapter 4 discusses how degrees of field extensions behave in towers, and studies the concept of an element in a field extension being algebraic over the base field. The notion of irreducibility of polynomials and the phenomenon of unique factorization in polynomial rings are studied in Chapter 5, and immediately after, the relation between the degree of the field generated by an element and the degree of its minimal polynomial is derived in Chapter 6. Finally, these results are put together in Chapter 7 to arrive at the algebraic criterion for the constructibility of a real number.

Although the treatment of divisibility in the integers in Chapter 1 is somewhat standard, the remaining chapters are a little less traditional. As described above, the goal is to get to constructibility with minimum fuss, but without sacrificing richness. For instance, in the chapter on rings and fields (Chapter 2), I discuss numerous examples of such objects and I discuss subrings generated by elements, but I avoid talking of ideals since I do not have a formal need for this concept. (On the other hand, in Chapter 6, the set denoted $I_{F,a}$ is after all just an ideal of $F[x]$, so I take advantage of this opportunity to give them a sequence of exercises concerning ideals in general.) Similarly, by working within a fixed field extension K/F, I never deal with the abstract construction of field extensions generated by roots of polynomials, and instead focus on field extensions generated by specific elements of the overfield K. (In fact, the issue of an element a of K being algebraic or transcendental over the subfield F is motivated by the question of when the field generated over F by a equals the ring generated over F by a.)

Along the way, I have tried to develop topics that a high-school mathematics teacher might find interesting. For instance, in the chapter on divisibility in the integers (Chapter 1), I include problems that show the validity of various divisibility tests (such as tests of divisibility by 3 and divisibility by 11). In the same chapter, I include a discussion on the Euclidean algorithm for finding the greatest common divisor of two integers, following an exercise where one has to show that if $a = bq + r$, then $\gcd(a, b) = \gcd(b, r)$. In Chapter 2, I include a problem that shows in a series of steps that unique factorization fails in a very natural "number system," an exercise that I hope will help the reader appreciate the significance of unique prime factorization in the integers. In Chapter 4, I introduce the concept of algebraic and transcendental numbers, and then discuss the transcendentality of certain specific numbers. I include a problem on showing that e is irrational, and in the notes to this same chapter, I explain why there are "so many more" transcendental numbers than there are algebraic numbers. In the chapter on polynomials (Chapter 5), I include discussions on the Fundamental Theorem of Algebra and on roots of polynomials. In the exercises to this chapter, I include problems that show why complex nonreal roots of polynomial equations with real coefficients come in pairs, why polynomials of odd degree with coefficients in the reals have a root in the reals, how the coefficients of a polynomial are related to its roots, how to obtain all n nth roots of a complex number given any one root, why the Lagrange interpolation polynomial is unique, and why synthetic division works the way it does. As well, in the notes to this chapter, I include discussions on the general problem of solving polynomial equations by radicals, and I outline Cardano's solution of the cubic.

A few words about the notes and the exercises. First the notes—they are meant to be *informal.* They started off as a vehicle by which I could try to provide glimpses into more advanced areas of mathematics as well as a vehicle by which I could communicate some of my own excitement about these areas. Very soon, however, they developed into a convenient receptacle for all sorts of remarks that I wanted to make, remarks that I felt should not be made in the text either for fear of derailing the course or for fear of giving away too much too soon. As such, one will find in the notes, besides pointers to theories beyond the scope of this book, comments on certain definitions, notes on certain proofs, remarks on specific examples, as well as occassional hints to some of the questions I ask in the text.

Now for the exercises. I have already noted above that many of the questions I ask within some of the examples I develop can serve as formal assignments. (To take an instance at random, in Examples 3.11 in Chapter 3, the questions asked in Examples 3.11.1, 3.11.2, 3.11.7, and 3.11.8 can all be assigned formally as problems.) For the most part, such problems assigned from the various examples I develop will be of a routine nature, designed to build familiarity with the material. As for the exercises at the end of each chapter, I have attempted to make many of them of some substance, exercises from which students will hopefully learn some significant mathematics. Of course, there is always a danger with such a philosophy in a beginning class, the danger that this approach may be too difficult for the students. To mitigate this somewhat, I have broken up many exercises into several digestible chunks, and I have provided copious hints. (I suspect that "guided discovery" is the surest way to learn mathematics, even when the students are seeing abstract mathematics for the first time.)

I believe that this book would serve very well as a gentle one-semester introduction to abstract algebra, after the students have had a basic introduction to sets and functions such as the introduction one gets in a typical undergraduate "discrete mathematics" course. By dwelling just a bit on the chapter on polynomials, working out all the exercises therein, a course taught out of this book would additionally provide some insight into what used to go by the name of "Theory of Equations." Such insight would be particularly useful to anybody teaching high-school mathematics. This book could also be used by anybody learning algebra on their own; the focus on exposition is designed to facilitate self-study.

Several colleagues have been of enormous help to me during the writing of this book. Pat Morandi was very encouraging about the worth of the project, and being a fellow-author, listened sympathetically to my travails. Besides, he put up with endless discussions on things like the definition of the greatest common divisor, when he would much rather be having endless discussions on things like the definition of étale cohomology. Also, he bravely volunteered to teach out of this book while on sabbatical at Indiana University. Jerry Gold, another brave soul, agreed to teach out of this book at California State University, Northridge, and provided several very valuable suggestions for improvement based on his experience. Ann Watkins read through portions of the book and made numerous comments that were extremely perceptive. She and Reinhard Laubenbacher both introduced me to the mechanics and the culture behind book publishing.

Many friends and family members helped as well. My brother Ananth and sister-in-law Vidya read through the preliminary version of the first few chapters and provided several suggestions. So did my college buddies KP and Shanks, as well as their spouses Malathi and Brinda. My good friends Henri and Stan read through the Introduction, and insisted that I retain the reference to the equitable distribution of pastry.

And of course, some of the most helpful individuals were the students in the Foundations of Algebra course at California State

University, Northridge. They are unfortunately too numerous to mention by name, but all these students should know that they were a joy to teach, and that it was they who were the *fundamental* reason why I wrote this book. It gives me particular pleasure to note that most of them are now established teachers themselves.

The National Science Foundation, as well as the Office of Research and Sponsored Projects and the College of Science and Mathematics at California State University, Northridge, provided generous support while this book was being written.

To all these people and organizations, I am very grateful.

B.A. Sethuraman
August 1996

Contents

Introduction

Most of us are introduced to number systems very early in our lives, when we first learn how to count. We begin by learning to add, using the numbers 1, 2, 3, ... Then, we learn about the process of "taking away," that is, the process of subtracting one number from another, and as a consequence, we learn about the number 0 as well as the numbers $-1, -2, -3, \ldots$. Thus acquainted with the integers, we learn multiplication as a shortcut to addition—adding four 3s together is the same as multiplying 3 by 4. After several years of multiplication tables, we are taught fractions (usually in the context of dividing two pies among five people), and as a result, we learn about the rational numbers.

Our introduction to the real numbers comes to us from two sources. On the one hand, we learn about square roots and cube roots, and are thus introduced to numbers like $\sqrt{2}$ and $\sqrt[3]{3}$. On the other hand, we learn from geometry the concept of the length of a line. We are told that real numbers are the numbers that correspond to lengths of line segments, that is, to points on the number line.

Finally, it is pointed out to us that although it seems as if only positive numbers can have square roots, this is in fact not true. The number i is introduced to us as the square root of -1, and we are told that from this, we get a new system of numbers (the complex numbers) by considering all expressions of the form $a + ib$, where a and b are real numbers. We learn to add, subtract, multiply, and divide with these numbers. We learn about the geometric interpretation of complex numbers and about de Moivre's theorem, and we are told that at least in theory, we can solve any polynomial equation over the complex numbers.

Needless to say, in spite of our developing great mechanical facility with the complex numbers, they remain a mystery to most of us. Somehow, it still does not seem *correct* that a negative number could have a square root! Merely *defining* i to be the square root of -1 seems rather contrived, yet these abstract expressions of the form $a + ib$ indeed seem to give us a set of numbers with wonderful properties.

The complex numbers are not the only numbers that we wonder about. At some point, we all wonder about even the most basic of numbers, the positive integers. They have endless fascination for us, and there is a wealth of questions that we ask ourselves about these numbers. (Some of these, such as Goldbach's conjecture, that every positive even integer greater than 4 can be written as a sum of two odd primes, or the question of the existence of infinitely many "twin" primes, that is, primes that differ by two, remain unsettled to this day.) At other times, we wonder about the rationals, this process of forming fractions, that seems to take care of dividing objects into equal parts. We wonder about the other numbers on the real line, how some of them have decimal expansions that go on forever without any repetition, and how they can all be approximated arbitrarily closely by rational numbers. And of course, we continue to wonder about this mysterious square root of -1.

It is precisely this wonder about numbers that has been responsible for the development of much of the mathematics of the last two centuries, and in particular, of what is often referred to today as "abstract" algebra. The attempt to understand the structure of these numbers and to solve some of the outstanding problems concerning them has led to the introduction of some very deep concepts. These concepts have in turn shed light on other areas of mathematics, as well as on areas of science and engineering, and have thus considerably enriched human knowledge.

One of the problems that these concepts have solved is that of constructibility. This is a problem that had baffled the Greeks, and had remained unsolved for about two thousand years: Can arbitrary geometric figures be constructed using just a straightedge and a compass? The most famous version of this question asks whether it is possible to trisect an arbitrary angle, that is, whether it is possible to construct an angle whose measure is one-third that of any given angle, using just a straightedge and a compass. As it turns out, a complete answer can be given to this question using only introductory algebraic concepts.

Our goal in this book will be to learn these introductory concepts and then apply them to the solution of the constructibility problem.

Our path will take us through rings, fields, and vector spaces. We will learn about field extensions and learn to differentiate between algebraic and transcendental numbers. We will study the division algorithm for polynomials and the notion of an irreducible polynomial, and we will realize that these concepts are exact analogs of the corresponding division algorithm for integers and the notion of a prime integer. (In fact, we will start our studies by examining divisibility and primes in the integers.) We will learn about the degree of a field extension, and we will relate this degree to dimensions of certain vector spaces. Finally, we will see how this degree affects constructibility.

A thorough understanding of these introductory concepts will enable you to proceed further into mathematics and understand some of the questions we have described above. For instance, a more advanced course will detail the algebra behind the process by which the complex numbers are formed from the reals and will discuss the concept of an ordered field. This will hopefully settle your confusion about negative numbers having square roots, and you will hopefully see that the formation of the complex numbers from the reals is really not the contrived process it first seems, but is instead something very natural. Similarly, a more advanced course that includes real analysis (or "advanced calculus") will illumine the relationship of the rationals to the reals. As a result of such a course, you will hopefully realize that the real numbers are precisely the numbers that arise when one tries to come to grips with the concept of decimal expansions that go on forever without repetition, and you will hopefully understand that it is perfectly natural that the rationals be *dense* in the reals, that is, that every real number be approximated arbitrarily closely by rational numbers.

How should you read this book? The answer, which applies to every book on mathematics, can be given in one word—*actively*. You may have heard this before, but it can never be overstressed—*you can only learn mathematics by doing mathematics*. This means much more than attempting all the problems assigned to you (although attempting every problem assigned to you is a *must*). What it means is that you should take time out to think through every sentence and confirm every assertion made. You should accept nothing on trust; instead, not only should you check every statement, you should also attempt to go beyond what is stated, searching for patterns, looking for connections with other material that you may have studied, and probing for possible generalizations.

Let us consider an example. On page 34 in Chapter 2, you will find the following sentence:

> Yet, even in this extremely familiar number system, multiplication is not commutative; for instance,
>
> $$\begin{pmatrix} 1 & 0 \\ 0 & 0 \end{pmatrix} \cdot \begin{pmatrix} 0 & 1 \\ 0 & 0 \end{pmatrix} \neq \begin{pmatrix} 0 & 1 \\ 0 & 0 \end{pmatrix} \cdot \begin{pmatrix} 1 & 0 \\ 0 & 0 \end{pmatrix}.$$

(The "number system" referred to is the set of 2×2 matrices whose entries are real numbers.) When you read a sentence such as this, the first thing that you should do is *verify the computation yourselves*. Mathematical insight comes from mathematical experience, and you cannot expect to gain mathematical experience if you merely accept somebody else's word that the product on the left side of the equation does not equal the product on the right side.

The very process of multiplying out these matrices will make the set of 2×2 matrices a more familiar system of objects, but as you do the calculations, more things can happen if you keep your eyes and ears open. Some or all of the following may occur:

1. You may notice that not only are the two products not the same, but that the product on the right side gives you the *zero* matrix. This should make you realize that although it may seem impossible that two nonzero "numbers" can multiply out to zero, this is only because you are confining your thinking to the real or

complex numbers. Already, the set of 2×2 matrices (with which you have at least some familiarity) contains nonzero elements whose product is zero.

2. Intrigued by this, you may want to discover other pairs of nonzero matrices that multiply out to zero. You will do this by taking arbitrary pairs of matrices and determining their product. It is quite probable that you will not find an appropriate pair. At this point you may be tempted to give up. However, you should not. You should try to be creative, and study how the entries in the various pairs of matrices you have selected affect the product. It may be possible for you to change one or two entries in such a way that the product comes out to be zero. For instance, suppose you consider the product

$$\begin{pmatrix} 1 & 1 \\ 1 & 1 \end{pmatrix} \cdot \begin{pmatrix} 4 & 0 \\ 2 & 0 \end{pmatrix} = \begin{pmatrix} 6 & 0 \\ 6 & 0 \end{pmatrix}$$

You should observe that no matter what the entries of the first matrix are, the product will always have zeros in the $(1,2)$ and the $(2,2)$ slots. This gives you some freedom to try to adjust the entries of the first matrix so that the $(1,1)$ and the $(2,1)$ slots also come out to be zero. After some experimentation, you should be able to do this.

3. You may notice a pattern in the two matrices that appear in our inequality on page 4. Both matrices have only one nonzero entry, and that entry is a 1. Of course, the 1 occurs in different slots in the two matrices. You may wonder what sorts of products occur if you take similar pairs of matrices, but with the nonzero 1 occuring at other locations. To settle your curiosity, you will multiply out pairs of such matrices, such as

$$\begin{pmatrix} 0 & 0 \\ 1 & 0 \end{pmatrix} \cdot \begin{pmatrix} 0 & 1 \\ 0 & 0 \end{pmatrix},$$

or

$$\begin{pmatrix} 0 & 0 \\ 1 & 0 \end{pmatrix} \cdot \begin{pmatrix} 0 & 0 \\ 1 & 0 \end{pmatrix}.$$

You will try to discern a pattern behind how such matrices multiply. To help you describe this pattern, you will let $e_{i,j}$ stand for the matrix with 1 in the (i,j)-th slot and zeros everywhere else, and you will try to discover a formula for the product of $e_{i,j}$ and $e_{k,l}$, where i, j, k, and l can each be any element of the set $\{1,2\}$.

4. You may wonder whether the fact that we considered only 2×2 matrices is significant when considering noncommutative multiplication or when considering the phenomenon of two nonzero elements that multiply out to zero. You will ask yourselves whether the same phenomena occur in the set of 3×3 matrices or 4×4 matrices. You will next ask yourselves whether they occur in the set of $n \times n$ matrices, where n is arbitrary. But you will caution yourselves about letting n be too arbitrary. Clearly n needs to be a positive integer, since "$n \times n$ matrices" is meaningless otherwise, but you will wonder whether n can be allowed to equal 1 if you want such phenomena to occur.

5. You may combine 3 and 4 above, and try to define the matrices $e_{i,j}$ analogously in the general context of $n \times n$ matrices. You will study the product of such matrices in this general context and try to discover a formula for their product.

Notice that a single sentence can lead to an enormous amount of mathematical activity! Every step requires you to be alert and actively involved in what you are doing. You observe patterns for yourselves, you ask yourselves questions, and you try to answer these questions on your own. In the process, you discover most of the mathematics yourselves. This is really the only way to learn mathematics (and in particular, it is the way every professional mathematician has learned the subject). Mathematical concepts are developed precisely because mathematicians observe patterns in various mathematical objects (such as the 2×2 matrices), and to have a good understanding of these concepts you must try to notice these patterns for yourselves.

To help you along, brief notes for each chapter have been included. These notes contain hints to some of the questions asked in the chapter, as well as general comments about some of the definitions, examples, and theories presented in the chapter. Do not rush to read these notes; you need to think independently about the material first!

Besides the willingness to read this book actively, the prerequisites for this book are small. You are expected to have some familiarity with the integers, as well as with the rationals, real and complex numbers, polynomials, and matrices. It would be helpful to be able to do proofs by induction. A rudimentary knowledge of set theory is assumed.

Exercises

1. Carry out the program in steps (1) through (5) above.

Divisibility in the Integers

We will begin our study with a very concrete set of objects, the integers, that is, the set $\{0, 1, -1, 2, -2, \ldots\}$. This set is traditionally denoted $\mathbb{Z}$ and is very familiar to us—in fact, we were introduced to this set so early in our lives that we think of ourselves as having grown up with the integers! Moreover, we view ourselves as having completely absorbed the process of integer division; we unhesitatingly describe 3 as dividing 99 and equally unhesitatingly describe 5 as *not* dividing 101.

As it turns out, this very familiar set of objects has an immense amount of structure to it. It turns out, for instance, that there are certain distinguished integers (the primes) that serve as building blocks for all other integers. These primes are rather beguiling objects; their existence has been known for over two thousand years, yet there are still several unanswered questions about them. They serve as building blocks in the following sense: every positive integer greater than 1 can be expressed uniquely as a product of primes. (Negative integers less than -1 also factor into a product of primes, except that they have a minus sign in front of the product.)

The fact that nearly every integer breaks up uniquely into building blocks is an amazing one; this is a property that holds in very few number systems. (In the exercises to Chapter 2 we will see an example of a number system whose elements *do not* factor uniquely into building blocks. Chapter 2 will also contain a discussion of what a "number system" is—see Remark 2.5.) On the other hand, there are some number systems where such a property does hold, notably polynomials, and we will find that the fact that polynomials also break up uniquely into building blocks is crucial to our treatment of constructibility. We will study polynomials in Chapter 5.

Our goal in this chapter is to prove that integers can be factored uniquely into primes. We will begin by examining the notion of divisibility and defining divisors and multiples. We will study the division algorithm and how it follows from the Well Ordering Principle. We will explore greatest common divisors and the notion of relative primeness. We will then introduce primes and prove our factorization theorem. Finally, we will look at what is widely considered as the ultimate illustration of the elegance of pure mathematics—Euclid's proof that there are infinitely many primes.

Let us start with something that *seems* very innocuous, but is actually rather profound. Write $\mathbb{N}$ for the nonnegative integers that is, $\mathbb{N} = \{0, 1, 2, 3, \ldots\}$. ($\mathbb{N}$ stands for "natural numbers," as the nonnegative integers are sometimes referred to.) Let S be any *nonempty* subset of $\mathbb{N}$. For example, S could be the set $\{0, 5, 10, 15, \ldots\}$, or the set $\{1, 4, 9, 16, \ldots\}$, or else the set $\{100, 1000\}$. The following is rather obvious: there is an element in S that is smaller than every other element in S, that is, S has a *smallest* or *least* element. This fact, namely that every nonempty subset of $\mathbb{N}$ has a least element, turns out to be a *crucial* reason why the integers possess all the other beautiful properties (such as a notion of divisibility, and the existence of prime factorizations) that make them so interesting.

Contrast the integers with another very familiar number system, the rationals, that is, the set $\{a/b \mid a \text{ and } b \text{ are integers, with } b \neq 0\}$. (This set is traditionally denoted by $\mathbb{Q}$.) Can you think of a nonempty subset of the positive rationals that fails to have a least element?

We will take this property of the integers as a fundamental *axiom*, that is, we will merely accept it as given and not try to prove it from more fundamental principles. Also, we will give it a name:

Well Ordering Principle: Every nonempty subset of the nonnegative integers has a least element.

Now let us look at divisibility. Why do we say that 2 divides 6? It is because there is another integer, namely 3, such that the product 2 times 3 *exactly* gives us 6. On the other hand, why do we say that 2 does not divide 7? This is because no matter how hard we search, we will not be able to find an integer b such that 2 times b equals 7. This idea will be the basis of our definition:

Definition 1.1

A (nonzero) integer d is said to *divide* an integer a (denoted $d|a$) if there exists an integer b such that $a = db$. If d divides a, then d is referred to as a *divisor* of a or a *factor* of a, and a is referred to as a *multiple* of d.

Observe that this is a slightly more general definition than most of us are used to—according to this definition, -2 divides 6 as well, since there exists an integer, namely -3, such that -2 times -3 equals 6. Similarly, 2 divides -6, since 2 times -3 equals -6. More generally, if d divides a, then all of the following are also true: $d|-a$, $-d|a$, $-d|-a$. (Take a minute to prove this formally!) It is quite reasonable to include negative integers in our concept of divisibility, but for convenience, we will often focus on the case where the divisor is positive.

The following easy result will be very useful:

Lemma 1.2

If d is a nonzero integer such that $d|a$ and $d|b$ for two integers a and b, then for any integers x and y, $d|(xa + yb)$. (In particular, $d|(a + b)$ and $d|(a - b)$.)

Proof Since $d|a$, $a = dm$ for some integer m. Similarly, $b = dn$ for some integer n. Hence $xa + yb = xdm + ydn = d(xm + yn)$. Since we have succeeded in writing $xa + yb$ as d times the integer $xm + yn$, we find that $d|(xa + yb)$. As for the statement in the parentheses, taking $x = 1$ and $y = 1$, we find that $d|a + b$, and taking $x = 1$ and $y = -1$, we find that $d|a - b$. □

The following lemma holds the key to the division process. Its statement is often referred to as the division algorithm. The Well Ordering Principle plays a central role in its proof.

Lemma 1.3 (Division Algorithm)
Given integers a and b with $b > 0$, there exist unique *integers q and r, with $0 \leq r < b$ such that $a = bq + r$.*

Remark 1.4
First, observe the range that r lies in. It is constrained to lie between 0 and $b - 1$ (with both 0 and $b - 1$ included as possible values for r). Next, observe that the lemma does not just state that integers q and r exist with $0 \leq r < b$ and $a = bq + r$, it goes further—it states that these integers q and r are *unique*. This means that if somehow one were to have $a = bq_1 + r_1$ and $a = bq_2 + r_2$ for integers q_1, r_1, q_2, and r_2 with $0 \leq r_1 < b$ and $0 \leq r_2 < b$, then q_1 must equal q_2 and r_1 must equal r_2. The integer q is referred to as the *quotient* and the integer r is referred to as the *remainder.*

Proof of Lemma 1.3 Let S be the set $\{a - bn \mid n \in \mathbb{Z}\}$. Thus, S contains the following integers: a $(= a - b \cdot 0)$, $a - b$, $a + b$, $a - 2b$, $a + 2b$, $a - 3b$, $a + 3b$, etc. Let S^* be the set of all those elements in S that are nonnegative, that is, $S^* = \{a - bn \mid n \in \mathbb{Z}, \text{ and } a - bn \geq 0\}$. It is not immediate that S^* is nonempty, but if we think a bit harder about this, it will be clear that S^* indeed has elements in it. For if a is nonnegative, then $a \in S^*$. If a is negative, then $a - b(a)$ is nonnegative (check!), so $a - b(a) \in S^*$. By the Well Ordering Principle, since S^* is a *nonempty* subset of $\mathbb{N}$, S^* has a least element; call it r. (The notation r is meant to be suggestive; this element will be the "r" guaranteed by the lemma.)

Since r is in S (actually in S^* as well), r must be expressible as $a - bq$ for some integer q, since *every* element of S is expressible as $a - bn$ for some integer n. (The notation q is also meant to be suggestive, this integer will be the "q" guaranteed by the lemma.) Since $r = a - bq$, we find $a = bq + r$. What we need to do now is to show that $0 \leq r < b$, and that q and r are unique.

Observe that since r is in S^* and since all elements of S^* are nonnegative, r must be nonnegative, that is $0 \leq r$. Now suppose $r \geq b$. We will arrive at a contradiction. Write $r = b + x$, where $x \geq 0$ (why is $x \geq 0$?). Writing $b + x$ for r in $a = bq + r$, we find $a = bq + b + x$, or $a = b(q + 1) + x$, or $x = a - b(q + 1)$. This form of x shows that x belongs to the set S (why?). Since we have already seen that $x \geq 0$, we find further that $x \in S^*$. But more is true: since

$x = r - b$ and $b > 0$, x must be less than r (why?). Thus, x is an element of S^* that is smaller that r—a contradiction to the fact that r is the least element of S^*! Hence, our assumption that $r \geq b$ must have been false, so $r < b$. Putting this together with the fact that $0 \leq r$, we find that $0 \leq r < b$, as desired.

Now for the uniqueness of q and r. Suppose $a = bq_1 + r_1$ and as well, $a = bq_2 + r_2$, for integers q_1, r_1, q_2, and r_2 with $0 \leq r_1 < b$ and $0 \leq r_2 < b$. Then $b(q_1 - q_2) = r_2 - r_1$. Thus, $r_2 - r_1$ is a multiple of b. Now the fact that $0 \leq r_1 < b$ and $0 \leq r_2 < b$ shows that $-b < r_2 - r_1 < b$. (Convince yourselves of this!) The only multiple of b in the range $(-b, b)$ (both endpoints of the range excluded) is 0. Hence, $r_2 - r_1$ must equal 0, that is, $r_2 = r_1$. It follows that $b(q_1 - q_2) = 0$, and since $b \neq 0$, we find that $q_1 = q_2$. □

Observe that to test whether a given (positive) integer d divides a given integer a, it is enough to write a as $dq + r$ $(0 \leq r < d)$ and examine whether r is zero or not. For $d|a$ if and only if the remainder obtained on dividing a by d (the integer r above) is zero.

Now, given two nonzero integers a and b, it is natural to wonder whether they have any divisors in common. Notice that 1 is automatically a common divisor of a and b, no matter what a and b are. Recall that $|a|$ denotes the absolute value of a, and notice that every divisor d of a is less than or equal to $|a|$. (Why?) Also, for every divisor d of a, we must have $d \geq -|a|$. (Why?) Similarly, every divisor d of b must be less than or equal to $|b|$ and greater than or equal to $-|b|$. It follows that every common divisor of a and b must be less than or equal to the lesser of $|a|$ and $|b|$, and must be greater than or equal to the greater of $-|a|$ and $-|b|$. Thus, there are only finitely many common divisors of a and b, and they all lie in the range $max(-|a|, -|b|)$ to $min(|a|, |b|)$.

We will now focus on a very special common divisor of a and b.

Definition 1.5

Given two (nonzero) integers a and b, the *greatest common divisor* of a and b (written as $\gcd(a, b)$) is the *largest* of the common divisors of a and b.

Note that since there are only finitely many common divisors of a and b, it makes sense to talk about the largest of the common divisors. (By contrast, must an infinite set of integers necessarily have a largest element? Must an infinite set of integers necessarily *fail* to have a largest element? What would your answers to these two questions be if we restricted our attention to an infinite set of positive integers?)

Notice that since 1 is already a common divisor, the greatest common divisor of a and b must be at least as large as 1. We can conclude from this that the greatest common divisor of two nonzero integers a and b must be *positive*.

If p and q are two positive integers and if q divides p, what must $gcd(p, q)$ be? Will your answer change if p is merely assumed to be a nonnegative integer, but q is still assumed to be a positive integer and still assumed to divide p?

Let us derive an alternative formulation for the greatest common divisor that will be very useful. Given two nonzero integers a and b, any integer that can be expressed in the form $xa + yb$ for some integers x and y is called a *linear combination* of a and b. (For example, $a = 1 \cdot a + 0 \cdot b$ is a linear combination of a and b; so are $3a - 5b$, $-6a + 10b$, $-b = 0 \cdot a + (-1) \cdot b$, etc.) Write P for the set of linear combinations of a and b that are *positive*. (For instance, if $a = 2$ and $b = 3$, then $-2 = (-1) \cdot 2 + (0) \cdot 3$ would not be in P as -2 is negative, but $7 = 2 \cdot 2 + 3$ would be in P as 7 is positive.) Now here is something remarkable: the smallest element in P turns out to be the greatest common divisor of a and b! We will prove this below.

Theorem 1.6

Given two nonzero integers a and b, let P be the set $\{xa + yb | x, y \in \mathbb{Z}, xa + yb > 0\}$. Let d be the least element in P. Then $d = gcd(a, b)$. Moreover, every element of P is divisible by d.

Proof First observe that P is not empty. For if $a > 0$, then $a \in P$, and if $a < 0$, then $-a \in P$. Thus, since P is a nonempty subset of $\mathbb{N}$ (actually, of the positive integers as well), the Well Ordering Principle guarantees that there is a least element d in P, as claimed in the statement of the theorem.

To show that $d = \gcd(a, b)$, we need to show that d is a common divisor of a and b, and that d is the largest of all the common divisors of a and b.

First, since $d \in P$, and since every element in P is a linear combination of a and b, d itself can be written as a linear combination of a and b. Thus, there exist integers x and y such that $d = xa + yb$. (Note: These integers x and y need not be unique. For instance, if $a = 4$ and $b = 6$, we can express 2 as both $(-1) \cdot 4 + 1 \cdot 6$ and $(-4) \cdot 4 + 3 \cdot 6$. However, this will not be a problem; we will simply pick one pair x, y for which $d = xa + yb$ and stick to it.)

Let us show that d is a common divisor of a and b. Write $a = dq + r$ for integers d and r with $0 \le r < d$ (division algorithm). We need to show that $r = 0$. Suppose to the contrary that $r > 0$. Write $r = a - dq$. Substituting $xa + yb$ for d, we find that $r = (1 - xq)a + (-yq)b$. Thus, r is a positive linear combination of a and b that is less than d—a contradiction, since d is the smallest positive linear combination of a and b. Hence r must be zero, that is, d must divide a. Similarly, one can prove that d divides b as well, so that d is indeed a common divisor of a and b.

Now let us show that d is the largest of the common divisors of a and b. This is the same as showing that if c is any common divisor of a and b, then c must be no larger than d. So let c be any common divisor of a and b. Then, by Lemma 1.2 and the fact that $d = xa + yb$, we find that $c|d$. Thus, $c \le |d|$ (why?). But since d is positive, $|d|$ is the same as d. Thus, $c \le d$, as desired.

To prove the last statement of the theorem, note that we have already proved that $d|a$ and $d|b$. By Lemma 1.2, d must divide all linear combinations of a and b, and must hence divide every element of P.

We have thus proved our theorem. □

In the course of proving Theorem 1.6 above, we have actually proved something else as well, which we will state as a separate corollary:

Corollary 1.7
Every common divisor of two nonzero integers a and b divides their greatest common divisor.

Proof As remarked above, the ideas behind the proof of this corollary are already contained in the proof of Theorem 1.6 above. We saw there that if c is any common divisor of a and b, then c must divide d, where d is the minimum of the set P defined in the statement of the theorem. But this, along with the other arguments in the proof of the theorem, showed that d must be the greatest common divisor of a and b. Thus, to say that c divides d is really to say that c divides the greatest common divisor of a and b, thus proving the corollary. □

Exercise 12 will yield yet another description of the greatest common divisor.

Given two nonzero integers a and b for which one can find integers x and y such that $xa + yb = 2$, can you conclude from Theorem 1.6 that $\gcd(a, b) = 2$? If not, why not? What, then, are the possible values of $\gcd(a, b)$? Now suppose there exist integers x' and y' such that $x'a + y'b = 1$. Can you conclude that $\gcd(a, b) = 1$? (See the notes on Page 27 *after* you have thought about these questions for at least a little bit yourselves!)

Given two nonzero integers a and b, we noted that 1 is a common divisor of a and b. In general, a and b could have other common divisors greater than 1, but in certain cases, it may turn out that the greatest common divisor of a and b is precisely 1. We give a special name to this:

Definition 1.8

Two nonzero integers a and b are said to be *relatively prime* if $\gcd(a, b) = 1$.

We immediately have the following:

Corollary 1.9

Given two nonzero integers a and b, $\gcd(a, b) = 1$ if and only if there exist integers x and y such that $xa + yb = 1$.

Proof You should be able to prove this yourselves! (See the questions two paragraphs above Definition 1.8.) □

The following lemma will be useful:

Lemma 1.10

If $a|bc$ and $\gcd(a, b) = 1$ (where a, b, and c are nonzero integers), then $a|c$.

Proof Since $\gcd(a, b) = 1$, Theorem 1.6 shows that there exist integers x and y such that $1 = xa + yb$. Multiplying by c, we find that $c = xac + ybc$. Since $a|a$ and $a|bc$, a must divide c by Lemma 1.2. □

We are now ready to introduce the notion of a prime!

Definition 1.11
An integer p greater than 1 is said to be *prime* if its only divisors are ± 1 and $\pm p$. (An integer greater than 1 that is not prime is said to be *composite.*)

The first ten primes are 2, 3, 5, 7, 11, 13, 17, 19, 23, and 29. The hundredth prime is 541.

Primes are intriguing things to study. On the one hand, they should be thought of as being *simple,* in the sense that their only positive divisors are 1 and themselves. (This is sometimes described by the statement "primes have no nontrivial divisors.") On the other hand, there is an immense number of questions about them that are still unanswered, or at best, only partially answered. For instance: is every even integer greater than 4 expressible as a sum of two primes? (We saw this question in the introduction as "Goldbach's conjecture." The answer is unknown.) Are there infinitely many twin primes? (We saw this question earlier too—the answer to this is also unknown.) Is there any pattern to the occurence of the primes among the integers? Here, some partial answers are known. There are arbitrarily large gaps between consecutive primes, that is, given any n, it is possible to find two consecutive primes that differ by at least n. (See Exercise 8.) It is known that for any $n > 1$, there is always a prime between n and $2n$. (It is unknown whether there is a prime between n^2 and $(n + 1)^2$, however!) It is known that as n becomes very large, the number of primes less than n is approximately $n/ln(n)$. (This is the celebrated *Prime Number Theorem.*) Also, it is known that given any arithmetic sequence a, $a + d$, $a + 2d$, $a + 3d$, ..., where a and d are nonzero integers with $\gcd(a, d) = 1$, infinitely many of the integers that appear in this sequence are primes!

Those of you who find this fascinating should delve deeper into number theory, which is the branch of mathematics that deals with such questions. It is a wonderful subject with hordes of problems that will seriously challenge your creative abilities! For now, we will content ourselves with proving the unique prime factorization property and the infinitude of primes already referred to at the beginning of this chapter.

The following lemmas will be needed:

Lemma 1.12
Let p be a prime and a an arbitary integer. Then either $p|a$ or else $gcd(p, a) = 1$.

Proof If p already divides a, we have nothing to prove, so let us assume that p does not divide a. We need to prove that $\gcd(p, a) = 1$. Since any common divisor of p and a must in particular divide p, and since the only positive divisors of p are 1 and p, the only possible positive common divisors of p and a are 1 and p. Now, if p were a common divisor of p and a, then p would in particular divide a, contrary to our assumption. Hence, the only common divisor of p and a is 1, that is, $\gcd(p, a) = 1$. □

Lemma 1.13
Let p be a prime. If $p|ab$ for two integers a and b, then either $p|a$ or else $p|b$.

Proof If p already divides a, we have nothing to prove, so let us assume that p does not divide a. Then by Lemma 1.13 above, $\gcd(p, a) = 1$. It now follows from Lemma 1.10 that $p|b$. □

We are ready to prove our factorization theorem!

Theorem 1.14 (Fundamental Theorem of Arithmetic)
Every positive integer greater than 1 can be factored into a product of primes. The primes that occur in any two factorizations are the same, except perhaps for the order in which they occur in the factorization.

Remark 1.15
The statement of this theorem has two parts to it. The first sentence is an *existence* statement—it asserts that for every positive integer greater than 1, a prime factorization *exists.* The second sentence is a *uniqueness* statement. It asserts that except for rearrangement, *there can only be one prime factorization.* To understand this second assertion a little better, consider the two factorizations of 12 as $12 = 3 \times 2 \times 2$, and $12 = 2 \times 3 \times 2$. The orders in which the 2's and the 3

appear are different, but in both factorizations, 2 appears twice, and 3 appears once. The uniqueness part of the theorem tells us that no matter how 12 is factored, we will at most be able to rearrange the order in which the two 2's and the 3 appear such as in the two factorizations above, but every factorization must consist of exactly two 2's and one 3.

Proof of Theorem 1.14 We will prove the existence part first. The proof is very simple. Given any integer a greater than 1, either it is prime or it is not. If it is prime, then "$a = a$" is its prime factorization. If it is not, a must factor as bc for suitable integers b and c, with $b < a$ and $c < a$. If b and c are themselves prime, then "$a = bc$" is the desired prime factorization. If not, either b or c is not prime (or perhaps both are not prime). If, say, b is not prime, then $b = de$, for suitable integers d and e with $d < b$ and $e < b$. At this stage, we have $a = bc = dec$. If all three of d, e, and c are prime, then "$a = dec$" is the desired prime factorization. If not, then one or more of the three integers d, e, and c must admit further factors...This process must eventually stop, since at each stage the factors are becoming smaller and smaller, and the smallest factor we are allowed to have at any stage is 2. When the process stops, we will have a factorization of a into primes!

Let us move on to the uniqueness part of the theorem. The basic ideas behind the proof of this portion of the theorem are quite simple as well. The key is to recognize that if an integer a has two prime factorizations, then some prime in the first factorization must equal some prime in the second factorization. This will then allow us to cancel primes pair by pair in the two factorizations and conclude that the two factorizations must be the same.

So suppose that a is some positive integer greater than 1, and suppose that we have two prime factorizations

$$a = p_1^{n_1} \cdots p_s^{n_s} = q_1^{m_1} \cdots q_t^{m_t},$$

where the p_i $(i = 1, \ldots, s)$ are distinct primes, and the q_j $(j = 1, \ldots, t)$ are distinct primes, and the n_i and the m_j are positive integers. (By "distinct primes" we mean that $p_1, p_2, \ldots, p_s$ are all different from one another, and similarly, $q_1, q_2, \ldots, q_t$ are all different from one another.) Without loss of generality, we may assume that $s \geq t$. (What does this statement mean? It means that if instead t were greater than s, then we could simply reverse the roles of the p's and the q's in the arguments below, and the proof would still work!) Since p_1 divides a, and since $a = q_1^{m_1} \cdots q_t^{m_t}$, p_1 must divide $q_1^{m_1} \cdots q_t^{m_t}$. Now, we proved in Lemma 1.13 that if a prime divides a product of two integers, then it must divide one of these two integers. This generalizes easily to the statement that if a prime divides a product of k integers ($k \geq 2$), then it must divide one of these k integers. (Exercise 3 asks you to prove this!) In our situation, since p_1 divides the product $q_1^{m_1} \cdots q_t^{m_t}$, it must divide one of the factors of this product, that is, it must divide one of the q_j. Relabeling the primes q_j if necessary (remember, we do not consider a rearrangement of primes to be a different factorization), we may assume that p_1 divides q_1. Since the only positive divisors of q_1 are 1 and q_1, we find $p_1 = q_1$.

Now that we have proved that $p_1 = q_1$, let us prove that the exponent n_1 of p_1 in the first factorization must equal the exponent m_1 of q_1 in the second factorization. For suppose that $n_1 > m_1$. Since we know $p_1 = q_1$, let us for convenience write the two factorizations as

$$a = p_1^{n_1} \cdots p_s^{n_s} = p_1^{m_1} q_2^{m_2} \cdots q_t^{m_t}.$$

Canceling off $p_1^{m_1}$ on both sides, we will find

$$a = p_1^{n_1 - m_1} \cdots p_s^{n_s} = q_2^{m_2} \cdots q_t^{m_t}.$$

Since $n_1 - m_1 > 0$ (by assumption), p_1 divides the left-hand side, and hence the right-hand side as well. Another application of Exercise 3 shows that p_1 must divide one of $q_2, \ldots, q_m$. Since $p_1 = q_1$, this means that q_1 must divide one of $q_2, \ldots, q_m$. As before, since $q_2, \ldots, q_m$, are all primes, if q_1 divides one of $q_2, \ldots, q_m$, then q_1 must actually *equal* one of $q_2, \ldots, q_m$. But this is a contradiction, since we assumed that $q_1, q_2, \ldots, q_m$ are all distinct. Hence our assumption that $n_1 > m_1$ is flawed. We can prove similarly (do so!) that m_1 cannot be greater than n_1 either, which shows that $n_1 = m_1$.

At this stage, we have $p_1 = q_1$ and $n_1 = m_1$. This means as well that $p_1^{n_1} = q_1^{m_1}$.

Note that if $s = 1$, then $t = 1$ as well by our assumption that $s \geq t$, and the two factorizations of a must have been $a = p_1^{n_1} = q_1^{m_1}$.

Since we have already proved that $p_1 = q_1$ and $n_1 = m_1$, we would have proved our theorem!

So suppose $s > 1$. Canceling off $p_1^{n_1}$ and $q_1^{m_1}$ from the two factorizations of a, we find

$$p_2^{n_2} \cdots p_s^{n_s} = q_2^{m_2} \cdots q_t^{m_t}.$$

Once again, p_2 must divide some prime on the right, and must therefore be equal to some prime on the right. (How do we know that there are any primes at all on the right at this stage, that is, how do we know that $t \neq 1$?) By relabeling the primes if necessary, we may assume that $p_2 = q_2$. As before, we can show that both $n_2 > m_2$ and $m_2 > n_2$ will give us contradictions, so we conclude that n_2 must equal m_2. So far, we have $p_1 = q_1$, $p_2 = q_2$, $n_1 = m_1$, $n_2 = m_2$. Canceling off $p_2^{n_2}$ from the left and $q_2^{m_2}$ from the right and proceeding similarly till there are no more primes left to cancel, we find that $s = t$ (why is $s > t$ not possible?), $p_1 = q_1, \ldots, p_s = q_s$, and $n_1 = m_1$, $\ldots, n_s = m_s$. Thus, except for rearrangement, the two factorizations of a are indeed the same! □

Remark 1.16

While Theorem 1.14 only talks about integers greater than 1, a similar result holds for integers less than -1 as well: every integer less than -1 can be factored as -1 times a product of primes. The primes that occur in any two factorizations are the same, except perhaps for the order in which they occur in the factorizations.

Remark 1.17

Suppose we have a relation $a = bc$ between three positive integers a, b, and c. Stringing together the prime factorizations of b and c, we get a factorization of bc into a product of primes. On the other hand, bc is just a, and a has its own prime factorization as well. By the uniqueness of prime factorizations, the prime factorization of bc that we get from stringing together the prime factorizations of b and c *must* be *the* prime factorization of a. (For example, if $b = 36 = 2^2 \cdot 3^2$ and $c = 15 = 3 \cdot 5$, then $2^2 \cdot 3^2 \cdot 3 \cdot 5$ is a prime factorization of the product $36 \cdot 15 = 540$, and by the uniqueness of prime factorization, $2^2 \cdot 3^2 \cdot 3 \cdot 5$ *must* be *the* prime factorization of 540. Of course, this factorization is more commonly written as $2^2 \cdot 3^3 \cdot 5$.) In particular, the prime factors of b (and c) must be a subset of the prime factors of a.

Now suppose that a prime p occurs to the power x in the factorization of a, to the power y in the factorization of b, and to the power z in the factorization of c. Stringing together the factorizations of b and c, we find that p occurs to the power $y + z$ in the factorization of bc. Since the factorization of bc is just the factorization of a and since p occurs to the power x in the factorization of a, we find that $x = y + z$. In particular, $y \leq x$. Together with our earlier observation, this shows that if a and b are positive integers with $b|a$, the prime factors of b must be a subset of the prime factors of a, and the exponent of any prime p in the prime factorization of b can be no larger than the exponent of p in the prime factorization of a. Conversely, if a and b are positive integers such that the prime factors of b are a subset of the prime factors of a, and the exponent of any prime factor p in the prime factorization of b is no larger than the exponent of p in the prime factorization of a, then it is easy to check (do so!) that $b|a$. These observations will be very useful, particularly in the exercises.

Having proved the Fundamental Theorem of Arithmetic, there remains the question of showing that there are infinitely many primes. (What is wrong with the following argument?—There are infinitely many positive integers. Each of them factors into primes by the theorem that we have just proved. Hence there must be infinitely many primes.) The proof that we provide is due to Euclid, and is justly celebrated for its beauty.

Theorem 1.18 (Euclid)
There exist infinitely many prime numbers.

Proof Assume to the contrary that there are only finitely many primes. Label them $p_1, p_2, \ldots, p_n$. (Thus, we assume that there are n primes.) Consider the integer $a = p_1p_2\cdots p_n + 1$. Since $a > 1$, a admits a prime factorization by Theorem 1.14. Let q be any prime factor of a. Since the set $\{p_1, p_2, \ldots, p_n\}$ contains all the primes, q must be in this set, so q must equal, say, p_i. But then, $a = q(p_1p_2\cdots p_{i-1}p_{i+1}\cdots p_n) + 1$, so we get a remainder of 1 when we divide a by q. In other words, q cannot divide a. This is a contradiction. Hence there must be infinitely many primes! □

Exercises

1. In this exercise, we will formally prove the validity of various quick tests for divisibility that we learn in high school!

   (a) Prove that an integer is divisible by 2 if and only if the digit in the units place is divisible by 2. (Hint: Look at a couple of examples: $58 = 5 \cdot 10 + 8$, while $57 = 5 \cdot 10 + 7$. What does Lemma 1.2 suggest in the context of these examples?)

   (b) Prove that an integer is divisible by 4 if and only if the integer represented by the tens digit and the units digit is divisible by 4. (To give you an example, the "integer represented by the tens digit and the units digit" of 1024 is 24, and the assertion is that 1024 is divisible by 4 if and only if 24 is divisible by 4—which it is!)

   (c) Prove that an integer is divisible by 8 if and only if the integer represented by the thousands digit and the tens digit and the units digit is divisible by 8.

   (d) Prove that an integer is divisible by 3 if and only if the sum of its digits is divisible by 3. (For instance, the sum of the digits of 1024 is $1 + 0 + 2 + 4 = 7$, and the assertion is that 1024 is divisible by 3 if and only if 7 is divisible by 3—and therefore, since 7 is not divisible by 3, we can conclude that 1024 is not divisible by 3 either! Here is a hint in the context of this example: $1024 = 1 \cdot 1000 + 0 \cdot 100 + 2 \cdot 10 + 4 = 1 \cdot (999 + 1) + 0 \cdot (99 + 1) + 2 \cdot (9 + 1) + 4$. What can you say about the terms containing 9, 99, and 999 as far as divisibility by 3 is concerned? Then, what does Lemma 1.2 suggest?)

   (e) Prove that an integer is divisible by 9 if and only if the sum of its digits is divisible by 9.

   (f) Prove that an integer is divisible by 11 if and only if the difference between the sum of the digits in the units place, the hundreds place, the ten thousands place, ... (the places corresponding to the even powers of 10) and the sum of the digits in the tens place, the thousands place, the hundred thousands place, ... (the places corresponding to the odd powers of 10) is divisible by 11. (Hint: $10 = 11 - 1$, $100 = 99 + 1$,

$1000 = 1001 - 1$, $10000 = 9999 + 1$, etc. What can you say about the integers 11, 99, 1001, 9999, etc. as far as divisibility by 11 is concerned?)

2. Given nonzero integers a and b, with $b > 0$, write $a = bq + r$ (division algorithm). Show that $\gcd(a, b) = \gcd(b, r)$.

   (This exercise forms the basis for the Euclidean algorithm for finding the greatest common divisor of two nonzero integers. For instance, how do we find the greatest common divisor of, say, 48 and 30 using this algorithm? We divide 48 by 30 and find a remainder of 18, then we divide 30 by 18 and find a remainder of 12, then we divide 18 by 12 and find a remainder of 6, and finally, we divide 12 by 6 and find a remainder of 0. Since 6 divides 12 evenly, we claim that $\gcd(48, 30) = 6$. What is the justification for this claim? Well, applying the statement of this exercise to the first division, we find that $\gcd(48, 30) = \gcd(30, 18)$. Applying the statement to the second division, we find that $\gcd(30, 18) = \gcd(18, 12)$. Applying the statement to the third division, we find that $\gcd(18, 12) = \gcd(12, 6)$. Since the fourth division shows that 6 divides 12 evenly, $\gcd(12, 6) = 6$. Working our way backwards, we obtain $\gcd(48, 30) = \gcd(30, 18) = \gcd(18, 12) = \gcd(12, 6) = 6$.)

3. Show using induction and Lemma 1.13 that if a prime p divides a product of integers $a_1 \cdot a_2 \cdots a_k$ ($k \geq 2$), then p must divide one of the a_i's.

4. Given nonzero integers a and b, let $h = a/\gcd(a, b)$ and $k = b/\gcd(a, b)$. Show that $\gcd(h, k) = 1$.

5. Show that if a and b are nonzero integers with $\gcd(a, b) = 1$, and if c is an arbitrary integer, then $a|c$ and $b|c$ together imply $ab|c$. Give a counterexample to show that this result is false if $\gcd(a, b) \neq 1$. (Hint: Just as in the proof of Lemma 1.10, use the fact that $\gcd(a, b) = 1$ to write $1 = xa + yb$ for suitable integers x and y, and then multiply both sides by c. Now stare hard at your equation!)

6. The *Fibonacci Sequence,* $1, 1, 2, 3, 5, 8, 13, \cdots$ is defined as follows: If a_i stands for the ith term of this sequence, then $a_1 = 1$, $a_2 = 1$, and for $n \geq 3$, a_n is given by the formula $a_n = a_{n-1} + a_{n-2}$. Prove that for all $n \geq 2$, $\gcd(a_n, a_{n-1}) = 1$.

7. Given an integer $n \geq 1$, recall that $n!$ is the product $1 \cdot 2 \cdot 3 \cdots (n-1) \cdot n$. Show that the integers $(n+1)! + 2, (n+1)! + 3, \ldots, (n+1)! + (n+1)$ are all composite.

8. Use Exercise 7 to prove that given any positive integer n, one can always find *consecutive* primes p and q such that $q - p \geq n$.

9. If m and n are odd integers, show that 8 divides $m^2 - n^2$.

10. Let $n = p_1^{n_1} p_2^{n_2} \cdots p_k^{n_k}$ be the prime factorization of a positive integer, where for each i from 1 to k, p_i is a prime, and $n_1 \geq 1$. Show that the positive divisors of n are all those integers whose prime factorizations are of the form $p_1^{e_1} p_2^{e_2} \cdots p_k^{e_k}$, where for $i = 1, \cdots, k$, e_i is in the range $0 \leq e_i \leq n_i$. (As an example, the positive divisors of $36 = 2^2 3^2$, are the integers $2^0 3^0$ ($= 1$), $2^1 3^0$, $2^2 3^0$, $2^0 3^1$, $2^1 3^1$, $2^2 3^1$, $2^0 3^2$, $2^1 3^2$, and $2^2 3^2$.)

11. Use Exercise 10 to show that the number of positive divisors of n is $(n_1 + 1)(n_2 + 1) \cdots (n_k + 1)$.

12. Let m and n be positive integers. By allowing the exponents in the prime factorizations of m and n to equal 0 if necessary, we may assume that $m = p_1^{m_1} p_2^{m_2} \cdots p_k^{m_k}$ and $n = p_1^{n_1} p_2^{n_2} \cdots p_k^{n_k}$, where for $i = 1, \cdots, k$, p_i is prime, $m_i \geq 0$, and $n_i \geq 0$. (For instance, we can rewrite the factorizations $84 = 2^2 \cdot 3 \cdot 7$ and $375 = 3 \cdot 5^3$ as $84 = 2^2 \cdot 3 \cdot 5^0 \cdot 7$ and $375 = 2^0 \cdot 3 \cdot 5^3 \cdot 7^0$.) For each i, let $d_i = min(m_i, n_i)$. Prove that $\gcd(m, n) = p_1^{d_1} p_2^{d_2} \cdots p_k^{d_k}$.

13. Given two (nonzero) integers a and b, the *least common multiple* of a and b (written as $\text{lcm}(a, b)$) is defined to be the smallest of all the positive common multiples of a and b.

    (a) Show that this definition makes sense, that is, show that the set of positive common multiples of a and b has a smallest element.

    (b) Retaining the notation of Exercise 12 above, let $l_i = max(m_i, n_i)$ $(i = 1, \ldots, k)$. Show that $\text{lcm}(m, n) = p_1^{l_1} p_2^{l_2} \cdots p_k^{l_k}$.

    (c) Use Exercise 12 and Part 13b above to show that $\text{lcm}(a, b) = ab/\gcd(a, b)$.

    (d) Conclude that if if $\gcd(a, b) = 1$, then $\text{lcm}(a, b) = ab$.

14. Let $a = p^n$, where p is a prime and n is a positive integer. Prove that the number of integers x such that $1 \leq x \leq a$ and $\gcd(x, a) = 1$ is $p^n - p^{n-1}$.

    (More generally, if a is *any* integer greater than 1, one can ask for the number of integers x such that $1 \leq x \leq a$ and $\gcd(x, a) = 1$. This number is denoted by $\phi(a)$, and is referred to as *Euler's ϕ-function*. It turns out that if a has the prime factorization $p_1^{m_1} p_2^{m_2} \cdots p_k^{m_k}$, then $\phi(a) = \phi(p_1^{m_1}) \cdot \phi(p_2^{m_2}) \cdot \ldots \cdot \phi(p_k^{m_k})$! Delightful as this statement is, we will not delve deeper into it in this book, but you are encouraged to read about it in any introductory textbook on number theory.)

15. The series $1 + 1/2 + 1/3 + \cdots$ is known as the *harmonic series*. This exercise concerns the partial sums (see below) of this series.

    (a) Fix an integer $n \geq 1$, and let S_n denote the set $\{1, 2, \ldots, n\}$ Let 2^t be the highest power of 2 that appears in S_n. Show that 2^t does not divide any element of S_n other than itself.

    (b) For any integer $n \geq 1$, the *nth partial sum* of the harmonic series is the sum of the first n terms of the series, that is, it is the number $1 + 1/2 + 1/3 + \cdots 1/n$. Show that if $n \geq 2$, the nth partial sum is *not* an integer as follows:

       i. Clearing denominators, show that the nth partial sum may be written as a/b, where $b = n!$ and $a = (2 \cdot 3 \cdots n) + (2 \cdot 4 \cdots n) + (2 \cdot 3 \cdot 5 \cdots n) + \cdots + (2 \cdot 3 \cdots n - 1)$.

       ii. Let S_n and 2^t be as in part 15a above. Also, let 2^m be the highest power of 2 that divides $n!$. Show that $m \geq t \geq 1$ and that $m \geq m - t + 1 \geq 1$.

       iii. Conclude from part 15(b)ii above that 2^{m-t+1} divides b.

       iv. Use part 15a to show that 2^{m-t+1} divides all the summands in the expression in part 15(b)i above for a except the term $(2 \cdot 3 \cdots 2^t - 1 \cdot 2^t + 1 \cdots n)$.

       v. Conclude that 2^{m-t+1} does not divide a.

       vi. Conclude that the nth partial sum is not an integer.

16. Fix an integer $n \geq 1$, and let S_n denote the set $\{1, 3, 5, \ldots, 2n-1\}$. Let 3^t be the highest power of 3 that appears in S_n. Show that 3^t does not divide any element of S_n other than itself. Can you use

this result to show that the nth partial sums ($n \geq 2$) of a series analogous to the harmonic series (see Exercise 15 above) are not integers?

17. Prove using the unique prime factorization theorem that $\sqrt{2}$ is not a rational number. Using essentially the same ideas, show that $\sqrt{p}$ is not a rational number for any prime p. (Hint: Suppose that $\sqrt{2} = a/b$ for some two integers a and b with $b \neq 0$. Rewrite this as $a^2 = 2b^2$. What can you say about the exponent of 2 in the prime factorizations of a^2 and $2b^2$?)

Notes

Remarks on Theorem 1.6 It is very crucial that d be the *least* positive linear combination of a and b for you to be able to conclude that $\gcd(a, b) = d$. For instance, if you only know that there exist integers x and y such that $xa + yb = 2$, you cannot conclude that $\gcd(a, b) = 2$—for all you know, there may exist two other integers x' and y' such that $x'a + y'b = 1$!

Notice though that if you know that there exist integers x' and y' such that $x'a + y'b = 1$, you *can* conclude that $\gcd(a, b) = 1$. For 1 *has* to be the least positive linear combination of a and b, since there is no positive integer smaller than 1.

Remarks on the definition of the greatest common divisor We have defined the greatest common divisor of two nonzero integers a and b to be the largest of their common divisors (Definition 1.5), and we have noted that $\gcd(a, b)$ must be positive. On the other hand, Corollary 1.7 showed that every common divisor of a and b must divide $\gcd(a, b)$. Putting these together, we find that $\gcd(a, b)$ has the following specific properties:

1. $\gcd(a, b)$ is a positive integer.
2. $\gcd(a, b)$ is a common divisor of a and b.
3. Every common divisor of a and b must divide $\gcd(a, b)$.

You will find that many textbooks have turned these properties around and *have used these properites to define the greatest common divisor!* Thus, these textbooks define the greatest common divisor of a and b to be that integer d which has the following properties:

1. d is a positive integer.
2. d is a common divisor of a and b.
3. Every common divisor of a and b must divide d.

Of course, it is not immediately clear that such an integer d must exist, nor is it clear that it must be unique, and these books then give a proof of the existence and uniqueness of such a d. Furthermore, it is not immediately clear that the integer d yielded by this definition is the same as the greatest common divisor as we have defined it (although it will be clear if one takes a moment to think about it). The reason why many books prefer to define the greatest common divisor as above is that this definition applies (with a tiny modification) to other number systems where the concept of a "largest" common divisor may not exist. (In fact, we ourselves will give a similar definition of the greatest common divisor of two nonzero polynomials in Chapter 5—see Definition 5.9.) In the case of the integers, however, we prefer our Definition 1.5, since the largest of the common divisors of a and b is exactly what we would intuitively expect $\gcd(a, b)$ to be!

CHAPTER 2 Rings and Fields

In the previous chapter we studied the integers in detail, focusing on divisibility properties. Divisibility, of course, is defined via multiplication: we say d divides a if $a = db$ for some integer b. What we did not do in the last chapter is go deeper still—we did not analyze multiplication itself.

Abstract algebra begins with the observation that several sets that occur naturally in mathematics, such as the integers, the rationals, the set of 2×2 matrices with entries in the reals, the set of continuous functions from the reals to the reals, all come equipped with certain operations that allow one to combine any two elements of the set and come up with a third element. These operations go by different names, such as addition, multiplication, or composition (you would have seen the notion of composing two functions in calculus). Abstract algebra studies mathematics from the point of view of these operations, asking, for instance, what properties of a given mathematical set can be deduced just from the existence of a given operation on the set with a given list of properties. We will be dealing with some of the more rudimentary aspects of this approach to mathematics in this book.

However, do not let the abstract nature of the subject fool you into thinking that mathematics no longer deals with concrete objects! Abstraction grows only from extensive studies of the concrete, it is merely a device (albeit an extremely effective one) for codifying phenomena that simultaneously occur in several concrete mathematical sets. In particular, to understand an abstract concept well, you must work with the specific examples from which the abstract concept grew (remember the advice on active learning).

Let us look at $\mathbb{Z}$ again, focusing this time on the operations of addition and multiplication.

Given a set S, recall that a *binary operation* on S is a process that takes an ordered pair of elements from S and gives us a third member of the set. It is helpful to think of this in more abstract terms—a binary operation on S is just a function $f: S \times S \to S$, that is, a rule that assigns to each ordered pair (a, b), a third element $f(a, b)$. Given an aribitrary set S, it is quite easy to define binary operations on it, but it is much harder to define binary operations that satisfy additional properties. (See Exercise 1.) What will be crucial to us is that addition and multiplication are *special* binary operations on $\mathbb{Z}$ that satisfy certain extra properties.

Why are addition and multiplication binary operations? The process of adding two integers is of course familiar to us, but suppose we view addition abstractly as a rule that assigns to each ordered pair of integers (m, n) the integer $m + n$. (For instance, addition assigns to the ordered pair $(2, 3)$ the integer 5, to the ordered pair $(3, -4)$ the integer -1, to the ordered pair $(1, 0)$ the integer 1, etc.) It is clear then that addition is indeed a binary operation—it takes an ordered pair of integers, namely (m, n), and gives us a third integer, namely $m + n$. Similarly, multiplication too is a binary operation—it is a rule that assigns to every ordered pair of integers (m, n) the integer $m \cdot n$.

What are the properties of these binary operations? Let us consider addition first. It is customary to write $(\mathbb{Z}, +)$ to emphasize the fact that we are considering $\mathbb{Z}$ not just as a set of objects, but as a set with the binary operation of addition. (We will temporarily ignore the fact that $\mathbb{Z}$ has a second binary operation, namely multiplication, defined on it.) The first property that $(\mathbb{Z}, +)$ has is that $+$ is *associative*. That is, for all integers a, b, and c, $(a + b) + c = a + (b + c)$. The second property that $(\mathbb{Z}, +)$ has is the existence of an *identity element with respect to* $+$. This is the integer 0—it satisfies the condition $a + 0 = 0 + a = a$ all integers a. The third property of $(\mathbb{Z}, +)$ is the existence of *inverses with respect to* $+$. For every integer a, there is an integer b (depending on a) such that $a + b = b + a = 0$. (It is clear what this integer b is, it is just the integer $-a$.)

For those who have studied groups, this must sound familiar. What we notice is that the integers form a group with respect to addition. For those who have not studied groups, let us take a moment to introduce the concept. It turns out that the situation we have encountered above (namely, a set equipped with a binary operation with certain properties) arises in several different areas of mathematics. Precisely because the same situation appears in so many different contexts, it has been given a name and has been studied extensively as a subject in its own right.

Definition 2.1

A *group* is a set S with a binary operation $f: S \times S \to S$ such that

1. f is associative,
2. S has an identity element with respect to f, and
3. every element of S has an inverse with respect to f.

To emphasize that there are two ingredients in this definition—the set S and the operation f with these special properties—the group is sometimes written as (S, f), and S is often referred to as *a group with respect to the operation* f.

The reason that the integers form a group with respect to addition is that if we take the set "S" of this definition to be $\mathbb{Z}$, and if we take the binary operation "f" to be $+$, then the three conditions of the definition are met. There is a vast and beautiful theory about groups, which we will not cover in this book, but which you are encouraged to pursue on your own.

Observe that there is one more property of addition that we have not listed yet, namely *commutativity*. This is the property that for all integers a and b, $a + b = b + a$. In the language of group theory, this makes $(\mathbb{Z}, +)$ an abelian group:

Definition 2.2

An *abelian group* is one in which the function "f" in Definition 2.1 above satisfies the additional condition $f(a, b) = f(b, a)$ for all a and b in S.

Commutativity of addition is a crucial property of the integers; the only reason we delayed introducing it was to allow us first to introduce the notion of a group.

Now let us consider multiplication. As with addition, we write $(\mathbb{Z}, \cdot)$ to emphasize the fact that we are considering $\mathbb{Z}$ as a set with the binary operation of multiplication, temporarily ignoring the operation addition. As with addition, we find that multiplication is *associative*, that is, for all integers a, b, and c, $(a \cdot b) \cdot c = a \cdot (b \cdot c)$. Also, $\mathbb{Z}$ has an *identity with respect to multiplication*. This is the integer 1; it satisfies $a \cdot 1 = 1 \cdot a = a$ for all integers a.

Is $(\mathbb{Z}, \cdot)$ a group? That is, do the integers form a group with respect to multiplication? Check to see whether the three group axioms above hold for $(\mathbb{Z}, \cdot)$. What is the inverse with respect to multiplication of 1? Of 2? Of 0?

There are two more properties of multiplication we wish to consider. The first is that multiplication is *commutative*, that is, $a \cdot b = b \cdot a$ for all integers a and b. The second, which is not a property of just multiplication alone, but rather a property that connects multiplication and addition together, is the *distributivity of multiplication over addition*, that is, for all integers a, b, and c, $a \cdot (b + c) = a \cdot b + a \cdot c$.

There are other properties of these operations of course (for instance $a \cdot b = 0$ implies that either $a = 0$ or $b = 0$), but we will study these later. Let us meanwhile reflect on the properties that we have considered so far. Studying them closely, one gets the sense that these properties are somehow rather "natural." For instance, if one were to think of the integers as (intellectual) counting tools, then it is clear that addition must necessarily be commutative, since commutativity of addition corresponds to the fact that if you have a certain number of objects in one pile and a certain number in another, then the total number of objects can be obtained either by counting all the objects in the first pile and then all the objects in the second pile, or by counting all the objects in the second pile and then all the objects in the first pile.

This sense of these properties being "natural" is further reinforced when we consider other "number systems" that we encounter in mathematics. For instance, consider the set of all polynomials in one variable whose coefficients are real numbers, a set with which you undoubtedly are very familiar. (The real numbers are traditionally denoted by $\mathbb{R}$, and the set of all polynomials in one variable whose coefficients are real numbers is traditionally denoted by $\mathbb{R}[x]$.) This set, too, is more than just a collection of objects. Just as with the integers, $\mathbb{R}[x]$ has two binary operations, also called *addition* and *multiplication*. Given two polynomials $g(x) = \sum_{i=0}^{n} g_i x^i$ and $h(x) = \sum_{j=0}^{m} h_i x^i$, we add g and h by adding together the coefficients of the same powers of x, and we multiply g and h by multiplying each monomial $g_i x^i$ of g by each monomial $h_j x^j$ of h and adding the results together. (For instance, $(1 + x + x^2) + (x + \sqrt{3}x^3)$ is $1 + 2x + x^2 + \sqrt{3}x^3$, and $(1 + x + x^2) \cdot (x + \sqrt{3}x^3)$ is $x + x^2 + (1 + \sqrt{3})x^3 + \sqrt{3}x^4 + \sqrt{3}x^5$.) Furthermore, these binary operations on $\mathbb{R}[x]$ have the *same* properties as the corresponding operations on $\mathbb{Z}$.

It turns out that these properties of addition and multiplication are shared not just by $\mathbb{Z}$ and $\mathbb{R}[x]$, but by a whole host of "number systems" in mathematics. Because of the importance of such sets with two binary operations with these special properties, there is a special term for them—they are called *rings*.

Definition 2.3

A *ring* is a set R with two binary operations $+$ and $\cdot$ such that

1. $a + b = b + a$ for all elements a, b in R.
2. $a + (b + c) = (a + b) + c$ for all a, b, c in R.
3. There exists an element, denoted by 0, in R such that $a + 0 = a$ for all a in R.
4. For each a in R there exists an element, denoted by $-a$, in R such that $a + (-a) = 0$.
5. $a \cdot (b \cdot c) = (a \cdot b) \cdot c$ for all elements a, b, c in R.
6. There exists an element, denoted by 1, in R such that $a \cdot 1 = 1 \cdot a = a$ for all a in R.
7. $a \cdot (b + c) = a \cdot b + a \cdot c$ and $(a + b) \cdot c = a \cdot c + b \cdot c$ for all elements a, b, c in R.

Remark 2.4

The binary operation $+$ is usually referred to as *addition* and the binary operation $\cdot$ is usually referred to as *multiplication*, in keeping with the terminology for the integers and other familiar rings. As is the usual practice in high school algebra, one often suppresses the multiplication symbol, that is, one often writes ab for $a \cdot b$.

Remark 2.5
We have used the term "number system" at several places in the book without really being explicit about what a number system is. We did not have the language before this point to make our meaning precise, but what we had intended to convey loosely by this term is the concept of a set with two binary operations with properties much like those of the integers. But now that we have the language, let us be precise: a number system is just a ring as defined above!

It must be borne in mind however that "number system" is a nonstandard term: it is not used very widely, and when used at all, different authors mean different things by the term! So it is better to stick to "rings," which is standard.

Observe that we left out one important property of the integers in our definition of a ring, namely the commutativity of multiplication. And correspondingly, we have included both *left distributivity* ($a \cdot (b + c) = a \cdot b + a \cdot c$) and *right distributivity* ($(a + b) \cdot c = a \cdot c + b \cdot c$) of multiplication over addition. While this may seem strange at first, think about the set of 2×2 matrices with entries in $\mathbb{R}$. Convince yourselves that this is a ring with respect to the usual definitions of matrix addition and multiplication—see Example 2.7.6 ahead. Yet, even in this extremely familiar number system, multiplication is not commutative; for instance,

$$\begin{pmatrix} 1 & 0 \\ 0 & 0 \end{pmatrix} \cdot \begin{pmatrix} 0 & 1 \\ 0 & 0 \end{pmatrix} \neq \begin{pmatrix} 0 & 1 \\ 0 & 0 \end{pmatrix} \cdot \begin{pmatrix} 1 & 0 \\ 0 & 0 \end{pmatrix}.$$

Rings in which multiplication is not commutative are fairly common in mathematics, and hence requiring commutativity of multiplication in the definition of a ring would be too restrictive. On the other hand, there is no denying that a significant proportion of the rings that we come across indeed have multiplication that is commutative. Thus, it is reasonable to single them out as special cases of rings, and we have the following:

Definition 2.6
A *commutative ring* is a ring R in which $a \cdot b = b \cdot a$ for all a and b in R.

(Rings in which the multiplication is not commutative are refered to as *noncommutative* rings.)

Examples 2.7

The following are various examples of rings. (Once again, recall the advice in the Introduction on reading actively.)

1. The set of rational numbers, $\mathbb{Q}$, with the usual operations of addition and multiplication forms a ring. We know how to add and multiply two rationals very well (we hope!), and we *know* that all the ring axioms hold for the rationals. (One can take a more advanced perspective and *prove* that the ring axioms hold for the rationals, starting from the fact that they hold for the integers. Although sound, such an approach is unduly technical for a first course.) $\mathbb{Q}$ is, in fact, a commutative ring. $\mathbb{Q}$ has one crucial property (with respect to multiplication) that $\mathbb{Z}$ does not have. Can you discover it? (See the remarks on page 58 in the notes, but only after you have thought about this question on your own!)
2. In a like manner, both the reals, $\mathbb{R}$, and the complexes, usually denoted by $\mathbb{C}$, are rings. Again, we will not try to *prove* that the ring axioms hold; we will just invoke our intimate knowledge of $\mathbb{R}$ and $\mathbb{C}$ to recognize that they are rings.
3. The set of all real numbers of the form $a + b\sqrt{2}$, where a and b are arbitrary rational numbers, forms a ring. For instance, this includes numbers like $1/2 + 3\sqrt{2}$, $-1/7 + (1/5)\sqrt{2}$, etc. You *know* how to add and multiply two elements $a + b\sqrt{2}$ and $c + d\sqrt{2}$ of this set. But there is a subtle point here—are you sure that under this method of addition and multiplication, the sum and product of any two elements of this set also lie in this set? (Remember, a binary operation should take an ordered pair of elements to another element in the *same* set. If, say, the usual product of some two elements $a + b\sqrt{2}$ and $c + d\sqrt{2}$ of this set does not belong to this set, then our usual product will not be a valid binary operation on this set, and hence we cannot claim that this set is a ring!) Why do you think associativity of addition and multiplication and distributivity of multiplication over addition all follow from the fact that this set is contained in $\mathbb{R}$? Are all other ring axioms satisfied? You know that $\sqrt{2}$ is not a rational number (see Chapter 1, Exercise 17). Use this result to show that $a + b\sqrt{2} = 0$ if and only if both a and b are zero. (For reasons that will be explained in the next section, this ring is denoted by

$\mathbb{Q}[\sqrt{2}]$. See the notes on Page 58, but as always, *after* you have played with this example yourselves!)

4. The set of all complex numbers of the form $a + bi$, where a and b are arbitrary rational numbers and i stands for $\sqrt{-1}$, forms a ring. As with the previous example, you *know* how to multiply two elements $a + bi$ and $c + di$ of this set. Under this method of addition and multiplication, are the sum and product of any two elements of this set also in the set? How do associativity and distributivity follow from the fact that this set is contained in the complex numbers? Are all other ring axioms satisfied? Show that $a + bi = 0$ if and only if both a and b are zero. (See the notes on page 59 for a clue. For reasons that will be explained in the next section, this ring is denoted by $\mathbb{Q}[i]$.)
5. The set of rational numbers q that have the property that when q is written in the reduced form a/b with a, b integers and $\gcd(a, b) = 1$ the denominator b is *odd,* forms a ring. This set is usually denoted by $\mathbb{Z}_{(2)}$, and contains elements like $1/3$, $-5/7$, $6/19$, etc., but does not contain $1/4$ or $-5/62$. (Does $\mathbb{Z}_{(2)}$ contain $2/6$?) Strange as this ring may seem at first, it plays an important role in number theory. Notice that every element of $\mathbb{Z}_{(2)}$ is just a fraction (albeit of a particular kind). You know how to add and multiply two fractions together; couldn't you use this method to add and multiply any two elements of $\mathbb{Z}_{(2)}$? As with the previous two examples, what subtle point needs to be checked? What role does the fact that the denominators are odd play in ensuring that this subtle point is met? (The role of the odd denominators is rather crucial; make sure that you understand it!) Why do associativity and distributivity follow from the fact that $\mathbb{Z}_{(2)} \subseteq \mathbb{Q}$? Do the other ring axioms hold? Can you generalize this construction to other subsets of $\mathbb{Q}$ where the denominators have analogous properties? (See the notes on page 59 for some comments.)
6. The set of $n \times n$ matrices with entries in $\mathbb{R}$ ($M_n(\mathbb{R})$), where n is a positive integer, forms a ring with respect to the usual operations of matrix addition and multiplication. For almost all values of n, matrix multiplication is not commutative. (What is the exception?) Checking associativity of addition and multiplication and the distributivity of multiplication over addition is tedious, but you should check at least one of them so as to be familiar

with the process. (For example, try to prove that for any three matrices A, B, and C, $(A + B) + C = A + (B + C)$.) What is important is that you get a feel for how associativity and distributivity in $M_n(\mathbb{R})$ *derives* from the fact that associativity and distributivity hold for $\mathbb{R}$. What about the ring axioms other than associativity and distributivity? Do they hold? What are the additive and multiplicative identities? Would the ring axioms still be satisfied if we only considered the set of $n \times n$ matrices whose entries came from $\mathbb{Q}$? From $\mathbb{Z}$? Now suppose R is any ring. If we consider the set $M_n(R)$ of $n \times n$ matrices with entries in R with the usual definitions of matrix addition and multiplication, is $M_n(R)$ with these operations a ring? What if R is not commutative? Does this affect whether $M_n(R)$ is a ring or not? (See the notes on page 59 for some hints.)

7. $\mathbb{R}[x]$, the set of polynomials in one variable with coefficients from $\mathbb{R}$, forms a ring with respect to the usual operations of polynomial addition and multiplication. (We have considered this before.) Here, x denotes the variable. Of course, one could use *any* letter to represent the variable. For instance, one could refer to the variable as t, in which case the set of polynomials with coefficients in $\mathbb{R}$ would be denoted by $\mathbb{R}[t]$. Sometimes, to emphasize our choice of notation for the variable, we refer to $\mathbb{R}[x]$ as the set of polynomials *in the variable* x with coefficients in $\mathbb{R}$, and we refer to $\mathbb{R}[t]$ as the set of polynomials *in the variable* t with coefficients in $\mathbb{R}$. Both $\mathbb{R}[x]$ and $\mathbb{R}[t]$, of course, refer to the same set of objects. Likewise, we often write $f(x)$ (or $f(t)$) for a polynomial, rather than just "f," to emphasize that the variable is x (or t).

   If $f(x) = a_0 + a_1x + a_2x^2 + \cdots$ is a *nonzero* polynomial in $\mathbb{R}[x]$, the *degree* of $f(x)$ is the largest value of n for which $a_n \neq 0$, a_nx^n is known as the *highest term*, and a_n is known as the *highest coefficient*. Thus, the polynomials of degree 0 are precisely the constants. Polynomials of degree 1 are called *linear*, polynomials of degree 2 are called *quadratic*, polynomials of degree 3 are called *cubic*, and so on. *Note that we have not defined the degree of the zero polynomial.* This is on purpose—it will be convenient for the formulation of certain theorems if the zero polynomial does not have a degree!

It is worth recalling an elementary property of polynomials that we will use frequently: two polynomials are equal if and only if their coefficients are equal. That is, $\sum f_i x^i = \sum g_i x^i$ if and only if $f_i = g_i$ ($i = 0, 1, \ldots$). In particular, a polynomial $\sum f_i x^i$ equals 0 if and only if each $f_i = 0$.

Now just as with Example 2.7.6, try to prove that if f, g, and h are any three polynomials in $\mathbb{R}[x]$, then $(f + g) + h = f + (g + h)$. Your proof should invoke the fact that associativity holds in $\mathbb{R}$. Study what happens if we were to consider polynomials with coefficients from an arbitrary ring R, with the usual definition of addition and multiplication of polynomials. Do we still get a ring? Is this ring commutative? (See the notes on page 60 for some hints and more remarks.)

8. Here is a ring with only two elements! Divide the integers into two sets, the even integers and the odd integers. Let $[0]_2$ denote the set of even integers, and let $[1]_2$ denote the set of odd integers. Denote by $\mathbb{Z}/2\mathbb{Z}$ the set $\{[0]_2, [1]_2\}$. Of course, each element of this set is itself a set containing an infinite number of integers, but we will ignore this fact. Instead, we will think of $[0]_2$ and $[1]_2$ as just "numbers" in the new number system $\mathbb{Z}/2\mathbb{Z}$. In other words, we will view all the even integers together as one "number" of $\mathbb{Z}/2\mathbb{Z}$, and we will view all the odd integers together as another "number" of $\mathbb{Z}/2\mathbb{Z}$. How should we add and multiply these new numbers? Recall that if we add two even integers we get an even integer, if we add an even and an odd integer we get an odd integer, and if we add two odd integers we get an even integer. This suggests the addition rules in $\mathbb{Z}/2\mathbb{Z}$:

"+"	$[0]_2$	$[1]_2$
$[0]_2$	$[0]_2$	$[1]_2$
$[1]_2$	$[1]_2$	$[0]_2$

(There is an obvious way to interpret this table: if you want to know what "a" + "b" is, you go to the cell corresponding to row a and column b.) Similarly, we know that the product of two even integers is even, the product of an even integer and an odd integer is even, and the product of two odd integers is odd. This gives us the multiplication rules:

"·"	$[0]_2$	$[1]_2$
$[0]_2$	$[0]_2$	$[0]_2$
$[1]_2$	$[0]_2$	$[1]_2$

There is a formal process for proving that these definitions give us a new ring (that is, for proving that all the ring axioms hold), but we will not cover this in this book. You could certainly prove from first principles that the axioms are satisfied, but this is tedious. Instead, just accept the fact that we get a ring, and play with the ring to develop a feel for it. Now here is a nice exercise: How would you get a ring with three elements in it? Four?

9. Here is the answer to the previous two questions! Notice that the even integers are the ones that yield a remainder of 0 on dividing by 2, and the odd integers are the ones that yield a remander of 1. (These are the only two remainders possible when you divide by 2.) Now suppose you want a ring with three elements. The possible remainders when you divide by 3 are 0, 1, and 2. Write $[0]_3$ for the set of all those integers that yield a remainder of 0 when you divide them by 3. In other words, $[0]_3$ consists of all multiples of 3, that is, all integers of the form $3k$, $k \in \mathbb{Z}$. Write $[1]_3$ for the set of all those integers that yield a remainder of 1 (so $[1]_3$ consists of all integers of the form $3k + 1$, $k \in \mathbb{Z}$), and write $[2]_3$ for the set of all those integers that yield a remainder of 2 (so $[2]_3$ consists of all integers of the form $3k + 2$, $k \in \mathbb{Z}$). Write $\mathbb{Z}/3\mathbb{Z}$ for the set $\{[0]_3, [1]_3, [2]_3\}$. Just as in the case of $\mathbb{Z}/2\mathbb{Z}$, every element of this set is itself a set consisting of an infinite number of integers, but we will ignore this fact. How would you add two elements of this set? In $\mathbb{Z}/2\mathbb{Z}$, we defined addition using observations like "an odd integer plus an odd integer gives you an even integer." The corresponding observations here are "an integer of the form $3k + 1$ plus another integer of the form $3k + 1$ gives you an integer of the form $3k + 2$," "an integer of the form $3k + 1$ plus another integer of the form $3k + 2$ gives you an integer of the form $3k$," "an integer of the form $3k + 2$ plus another integer of the form $3k + 2$ gives you an integer of the form $3k + 1$," etc. We thus get the following addition table:

"+"	$[0]_3$	$[1]_3$	$[2]_3$
$[0]_3$	$[0]_3$	$[1]_3$	$[2]_3$
$[1]_3$	$[1]_3$	$[2]_3$	$[0]_3$
$[2]_3$	$[2]_3$	$[0]_3$	$[1]_3$

Similarly, study how the remainders work out when we multiply two integers. (For instance, we find that "an integer of the form $3k + 2$ times an integer of the form $3k + 2$ gives you an integer of the form $3k + 1$," etc.) Derive the following multiplication table:

"·"	$[0]_3$	$[1]_3$	$[2]_3$
$[0]_3$	$[0]_3$	$[0]_3$	$[0]_3$
$[1]_3$	$[0]_3$	$[1]_3$	$[2]_3$
$[2]_3$	$[0]_3$	$[2]_3$	$[1]_3$

This process can easily be generalized to yield a ring with n elements ($\mathbb{Z}/n\mathbb{Z}$) for any $n \geq 2$. Play with the case when $n = 4$.

10. Suppose R and S are two rings. (For example, take $R = \mathbb{Z}/2\mathbb{Z}$, and take $S = \mathbb{Z}/3\mathbb{Z}$.) Consider the Cartesian product $T = R \times S$, which is the set of ordered pairs (r, s) with $r \in R$ and $s \in S$. Define addition in T by $(r, s) + (r', s') = (r + r', s + s')$. Here, "$r + r'$" refers to the addition of two elements of R according to the defintion of addition in R, and similarly, "$s + s'$" refers to the addition of two elements of S according to the definition of addition in S. For instance, in $\mathbb{Z}/2\mathbb{Z} \times \mathbb{Z}/3\mathbb{Z}$, $([0]_2, [1]_3) + ([1]_2, [2]_3) = ([1]_2, [0]_3)$. Similarly, define multiplication in T by $(r, s) \cdot (r', s') = (r \cdot r', s \cdot s')$. Once again, "$r \cdot r'$" refers to the multiplication of two elements of R according to the definition of multiplication in R, and "$s \cdot s'$" refers to the multiplication of two elements of S according to the definition of multiplication in S. Thus, in $\mathbb{Z}/2\mathbb{Z} \times \mathbb{Z}/3\mathbb{Z}$ again, $([0]_2, [1]_3) \cdot ([1]_2, [2]_3) = ([0]_2, [2]_3)$. Verify that these definitions of addition and mulitplication indeed make T a ring. T is known as the *direct product* of R and S. What are the identity elements with respect to addition and multiplication? Now take $R = S = \mathbb{Z}$. Show that $T = \mathbb{Z} \times \mathbb{Z}$ contains several pairs of *nonzero* elements a and b such that $a \cdot b = 0$, even though $\mathbb{Z}$ itself does not contain such elements. Will this phenomenon continue to hold if R and S are arbitrary rings? (See the notes on page 61 for hints.)

Remark 2.8

The examples above should have convinced you that our definition of a ring (Definition 2.3 above) is rather natural, and that it very effectively models several number systems that arise in mathematics. Here is further evidence that our axioms are the "correct" ones. Notice that in all the rings that we have come across, the following properties hold:

1. The additive identity is unique, that is, there is precisely one element 0 in the ring that has the property that $a + 0 = a$ for all elements a in the ring.
2. The multiplicative identity is unique, that is, there is precisely one element 1 in the ring that has the property that $a \cdot 1 = 1 \cdot a = a$ for all elements a in the ring.
3. $a + b = a + c$ implies $b = c$ for any elements a, b, and c in the ring.
4. For every element a in the ring, there is precisely one element $-a$ that satisfies the condition that $a + (-a) = 0$.
5. For every element a in the ring, $-(-a)$ is just a.
6. $a \cdot 0 = 0 \cdot a = 0$ for all elements a.
7. $(-1) \cdot a = a \cdot (-1) = -a$ for all elements a.
8. More generally, $a \cdot (-b) = (-a) \cdot b = -(ab)$ for all elements a and b.
9. $(-1) \cdot (-1) = 1$.
10. More generally, $(-a) \cdot (-b) = ab$ for all elements a and b.

Now these properties all seem extremely natural, and we would certainly like them to hold in all rings. (More strongly, a ring in which any of these properties fail would appear very pathological to us!) Now, if our ring axioms were the "correct" ones, then the properties above would be deducible *from the ring axioms themselves,* thereby showing that they hold in all rings. As it turns out, this is indeed true: they *are* deducible from the axioms, and therefore, they *do* hold in every ring R. We will leave the verification of this as an exercise (see Exercise 2). Of course, it is not necessary for you to prove every property above, that would be very tedious. Just try to prove one or two to get a feel for how the properties follow from the axioms.

Property 3 above is known as *additive cancellation.* Notice that there is one property that is very similar to it, namely *multiplicative cancellation*: $a \cdot b = a \cdot c$ implies $b = c$, which *cannot* be deduced from the ring axioms. The reason that this cannot be deduced from the axioms is very simple: *multiplicative cancellation does not hold in all rings!* Can you think of an example of a ring R and elements a, b, and c in R such that $ab = ac$ yet $b \neq c$?

Subrings

In Examples 2.7.3, 2.7.4, and 2.7.5 above, we came across the following phenomenon: A ring R and a subset S of R that had the following two properties: For any s_1 and s_2 in S, $s_1 + s_2$ was in S and $s_1 s_2$ was in S. In Example 2.7.3, the ring R was $\mathbb{R}$, and the subset S was the set of all real numbers of the form $a + b\sqrt{2}$ with a and b rational numbers. In Example 2.7.4, R was $\mathbb{C}$ and S was the set of all complex numbers of the form $a + bi$ with a and b rational numbers. In Example 2.7.5, R was $\mathbb{Q}$, and S was the set of all reduced fractions with odd denominator. Moreover, in all three examples, we endowed S with binary operations in the following way: Given s_1 and s_2 in S, we viewed them as elements of R, and formed the sum $s_1 + s_2$ (the sum being defined according to the definition of addition in R). Next, we observed that $s_1 + s_2$ was actually in S (this is one of the two properties alluded to above). Similarly, we observed that $s_1 s_2$ (the product being formed according to the definition of multiplication in R) was also in S. These two facts hence gave us two binary operation on S. We then found that with respect to these binary operations, S was not just an arbitrary subset of R, it was actually a ring in its own right.

The crucial reason (although not the only reason) why the set S in all our examples was itself a ring was that S had the properties described at the beginning of the previous paragraph. We give these properties a name.

Definition 2.9
Given an arbitrary nonempty subset S of a ring R, we say that S is *closed under addition* if for any s_1 and s_2 in S, $s_1 + s_2$ is also in S. Similarly, we that S is *closed under multiplication* if for any s_1 and s_2 in S, $s_1 s_2$ is also in S.

As we have observed, if a subset S of a ring R is closed under addition, then the addition operation on R, when restricted to ordered pairs of elements of S, yields a binary operation on S (which we also call addition), and we say that the addition on S is *induced* by the addition on R. Similarly, when S is closed under multiplication, we get a binary operation on S (also called multiplication) that we say is *induced* by the multiplication on R.

Now suppose that S is a subset of a ring R that is closed with respect to addition and multiplication, and just as in our examples above, suppose that with respect to the induced operations, S is itself a ring. We will give a special name to this situation:

Definition 2.10
Let S be a subset of a ring R that is closed with respect to addition and multiplication. Suppose that $1 \in S$. Suppose further that with respect to these addition and multiplication operations on S that are induced from those on R, S is itself a ring. We say that S is a *subring* of R.

Examples 2.7.3, 2.7.4, and 2.7.5 above are therefore all instances of subrings: $\mathbb{Q}[\sqrt{2}]$ is a subring of $\mathbb{R}$, $\mathbb{Q}[i]$ is a subring of $\mathbb{C}$, and $\mathbb{Z}_{(2)}$ is a subring of $\mathbb{Q}$.

(See the notes on page 61 for a remark on Definition 2.10 above.)

Before we proceed to look at further examples of subrings, let us first consider a criterion that will help us decide whether a given subset of a ring is actually a subring.

Lemma 2.11
Let S be a subset of a ring R which has the following properties:
1. *S is closed under addition,*
2. *S is closed under multiplication,*
3. *1 is in S, and*
4. *For all $a \in S$, $-a$ is also in S.*

Then S is a subring of R.

Proof As discussed above, since S is closed with respect to addition and multiplication, the addition and multiplication operations on R induce addition and multiplication operations on S. Now consider addition. For any a, b, and c in S, we may view a, b, and c as elements of R, and since addition is associative in R, we find $(a + b) + c = a +$

$(b + c)$. Viewing a, b, and c back as elements of S in this equation, we find that the induced addition operation on S is associative. Similarly, since addition is commutative in R, the induced addition on S is commutative. Now we are given that $1 \in S$, so property (4) shows that -1 is also in S. From the fact that S is closed under addition, we find that $1 + (-1)$ is also in S, so 0 is in S. The relation $s + 0 = s$ holds for all $s \in S$, since it holds more generally for any $s \in R$. Thus, S has an additive identity, namely 0. For every $s \in S$, we are given that $-s$ is also in S, so every element of S has an additive inverse. As for multiplication, given a, b, and c in S, we may view these as elements of R, and since multiplication in R is associative, we find that $(ab)c = a(bc)$. As before, viewing a, b, and c back as elements of S in this equation, we find that the induced multiplication operation on S is associative. Since $s \cdot 1 = 1 \cdot s = s$ for all $s \in S$ (as this is true more generally for all $s \in R$), and since $1 \in S$, we find that S has a multiplicative identity, namely 1. Finally, exactly as in the arguments for associativity above, the relations $a(b + c) = ab + ac$ and $(a + b)c = ac + bc$ hold for all a, b, and c in S because they hold in R, so distributivity is satisfied. S is hence a ring in its own right with respect to the induced operations of addition and multiplication and it contains 1. Thus, S is a subring of R. □

The following are further examples of subrings. Play with these examples to gain familiarity with them. Check that they are indeed examples of subrings of the given rings by applying Lemma 2.11.

Examples 2.12

1. The set of all real numbers of the form $a + b\sqrt{2}$ where a and b are *integers* is a subring of $\mathbb{Q}[\sqrt{2}]$. Why? It is denoted by $\mathbb{Z}[\sqrt{2}]$.
2. The set of all complex numbers of the form $a + bi$ where a and b are integers is a subring of $\mathbb{Q}[i]$. It is denoted by $\mathbb{Z}[i]$. (It is often called the ring of *Gaussian integers.*)
3. Let $\mathbb{Z}[1/2]$ denote the set of all rational numbers that are such that when written in the reduced form a/b with $\gcd(a, b) = 1$, the denominator b is a power of 2. (Contrast this set with $\mathbb{Z}_{(2)}$.) This is a subring of $\mathbb{Q}$. What are the rational numbers that this ring has in common with $\mathbb{Z}_{(2)}$? (See the notes on page 61 for clues.)

4. The set of all real numbers of the form $a + b\sqrt{2} + c\sqrt{3} + d\sqrt{6}$, where a, b, c, and d are all rational numbers is a subring of the reals. Is the set of all real numbers of the form $a + b\sqrt{2} + c\sqrt{3}$ where a, b, and c are rationals a subring of the reals?
5. If S is a subring of a ring R, then $S[x]$ is a subring of $R[x]$. In particular, $\mathbb{Q}[x]$ is a subring of $\mathbb{R}[x]$, which in turn is a subring of $\mathbb{C}[x]$.
6. Similarly, If S is a subring of a ring R, then $M_n(S)$ is a subring of $M_n(R)$.
7. Let $U_n(\mathbb{R})$ denote the *upper triangular matrices,* that is, the subset of $M_n(\mathbb{R})$ consisting of all matrices whose entries below the main diagonal are all zero. Thus, $U_n(\mathbb{R})$ is the set of all $((a_{i,j}))$ in $M_n(\mathbb{R})$ with $a_{i,j} = 0$ for $i < j$. (You may have seen the notation "$((a_{i,j}))$" before: it denotes the matrix whose entry in the ith row and jth column is the element $a_{i,j}$.) Then $U_n(\mathbb{R})$ is a subring of $M_n(\mathbb{R})$. Why? For what values of n will $U_n(\mathbb{R})$ be the same as $M_n(\mathbb{R})$? Suppose we considered the set of *strictly upper triangular matrices,* namely the set of all $((a_{i,j}))$ in $M_n(\mathbb{R})$ with $a_{i,j} = 0$ for $i \leq j$. Would we still get a subring of $M_n(\mathbb{R})$?
8. Here is another subring of $M_n(\mathbb{R})$. For each real number r, let diag(r) denote the matrix in which each diagonal entry is just r and in which the off-diagonal entries are all zero. The set of matrices in $M_n(\mathbb{R})$ of the form diag(r) (as r ranges through $\mathbb{R}$) is then a subring. What observations can you make about the function from $\mathbb{R}$ to $M_n(\mathbb{R})$ that sends r to diag(r)? (See the notes on page 61.)

Subring Generated by an Element

We now consider a topic that will be quite essential to our study of field extensions in later chapters. This is the notion of the ring generated by a subring and an element. *We will consider only commutative rings,* even though the notion exists for noncommutative rings as well. Accordingly, let R be a commutative ring, and let S be a subring. (Must S be commutative as well?) Let a be any element in R. For instance, let R be the reals, and let S be the rationals. Let a be the real number $1 + \sqrt{2}$. In general, $S \cup \{a\}$ (the set consisting of a and all the elements of S) will not be a subring of R, since this new set may not be closed under addition and multiplication. For instance, $\mathbb{Q} \cup \{1+\sqrt{2}\}$ is not closed under either addition or multiplication. (The square of $1+\sqrt{2}$, which is $3+2\sqrt{2}$, is not in $\mathbb{Q} \cup \{1+\sqrt{2}\}$. Similarly, the sum of, say 2 and $1+\sqrt{2}$, which is $3+\sqrt{2}$ is not in $\mathbb{Q} \cup \{1+\sqrt{2}\}$. Again, the product of, say 3 and $1+\sqrt{2}$, which is $3+3\sqrt{2}$, is not in $\mathbb{Q} \cup \{1+\sqrt{2}\}$.) One could then ask: If in general $S \cup \{a\}$ is not a subring of R, what are the elements of R that you should adjoin to the set $S \cup \{a\}$ to get a set that is actually a subring of R?

In our example above, our set failed to have the element $(1+\sqrt{2})^2$. It similarly does not have $(1+\sqrt{2})^3$, $(1+\sqrt{2})^4, \ldots$. In the general setting, to get a subring of R that contains both S and a, it is clear that we need to be able to multiply a with itself any number of times, since our desired set must be closed under multiplication. Hence, we need to adjoin all the elements $a^2, a^3, \ldots$ Next, once all powers a^i are adjoined, we need to be able to multiply any power of a with any element of S, so we need to adjoin all products of the form sa^i, where s is an arbitrary element of S and a^i is an arbitrary power of a. (The assumption that R is commutative is being used here somewhere. Where exactly do you think it is used?) Once we have such products, we need to be able to add such products together if we are to have a ring (our target set must be closed under addition), so we need to have all elements of the form $s_0 + s_1 a + s_2 a^2 + \cdots + s_n a^n$, where the s_i are arbitrary elements of S, and $n \geq 0$. Is this enough? It turns out it is!

Definition 2.13

Let R be a commutative ring, S a subring, and a an element of R. An expression such as $s_0 + s_1 a + s_2 a^2 + \cdots + s_n a^n$ is called a *polynomial expression in a with coefficients in S*. Let $S[a]$ denote the set of all polynomial expressions in a with coefficients in S, that is, the set of all elements of R that can be written in the form $s_0 + s_1 a + s_2 a^2 + \cdots + s_n a^n$, for some $n \geq 0$, and some elements $s_0, s_1, \ldots, s_n$ in S. $S[a]$ is known as the *subring of R generated by S and a*. (If it is clear that we are working inside a fixed ring R, we often refer to $S[a]$ merely as the *ring generated by S and a*.)

Of course, we have yet to prove that $S[a]$ is actually a ring, but we will do so in a moment.

Notice that $S[a]$ includes both S and a. (Does it?) Our arguments above show that any subring of R that contains both S and a *must* contain all polynomial expressions in a with coefficients in S, that is, it *must* contain $S[a]$. $S[a]$ should thus be thought of as the *smallest* subring of R that contains both S and a.

Lemma 2.14

Let R be a commutative ring, and let S be a subring of R. Let a be an element of R. The set $S[a]$ defined above is a subring of R.

Proof Since $S \subset S[a]$, and since $1 \in S$, 1 is in $S[a]$. Every element in $S[a]$ is of the form $s_0 + s_1a + s_2a^2 + \cdots + s_na^n$ for some $n \geq 0$ and some elements $s_0, s_1, \ldots, s_n$ in S. The negative of such an element is $(-s_0) + (-s_1)a + (-s_2)a^2 + \cdots + (-s_n)a^n$, which is also a polynomial expression in a with coefficients in S, and is hence in $S[a]$. By Lemma 2.2.1, we only need to show that $S[a]$ is closed under addition and multiplication. You should be able to do this yourselves: show that the sum and product of two polynomial expressions in a with coefficients in S are also polynomial expressions in a with coefficients in S. (See the notes on page 62 for some clues.) □

Now let us consider some examples:

Examples 2.15

1. What, according to our definition above, is the subring of the reals generated by $\mathbb{Q}$ and $\sqrt{2}$? It is the set of all polynomial expressions in $\sqrt{2}$ with coefficients in $\mathbb{Q}$, that is, the set of all expressions of the form $q_0 + q_1\sqrt{2} + q_2(\sqrt{2})^2 + \cdots + q_n(\sqrt{2})^n$. Now let us look at these expressions more closely. Since $(\sqrt{2})^2 = 2$, $q_2(\sqrt{2})^2$ is just $2q_2$, $q_4(\sqrt{2})^4$ is just $4q_4$, etc. Similarly, $q_3(\sqrt{2})^3$ is just $2q_3\sqrt{2}$, $q_5(\sqrt{2})^5$ is just $4q_5\sqrt{2}$, etc. By collecting terms together, it follows that every polynomial expression in $\sqrt{2}$ with coefficients in $\mathbb{Q}$ can be written as $a + b\sqrt{2}$ for suitable rational numbers a and b. (For example, $1 + 2\sqrt{2} + (1/2)(\sqrt{2})^2 + (1/4)(\sqrt{2})^3$ can be rewritten as $2 + (5/2)\sqrt{2}$.) Hence, the subring of the reals generated by the rationals and $\sqrt{2}$ is the set of all real numbers of the form $a + b\sqrt{2}$.

It is for this reason that we denoted this ring $\mathbb{Q}[\sqrt{2}]$ as far back as Example 2.7.3.

2. Similarly, the subring of $\mathbb{Q}[\sqrt{2}]$ generated by $\mathbb{Z}$ and $\sqrt{2}$ is the set of all real numbers of the form $a + b\sqrt{2}$, where a and b are integers. This is why we denoted this ring $\mathbb{Z}[\sqrt{2}]$ in Example 2.12.1.
3. Using the fact that $i^2 = -1$, show that the subring of $\mathbb{C}$ generated by $\mathbb{Q}$ and i is the set of all complex numbers of the form $a + bi$, where a and b are rational numbers. This explains the notation $\mathbb{Q}[i]$ for the ring in Example 2.7.4.
4. Similarly, the subring of $\mathbb{Q}[i]$ generated by $\mathbb{Z}$ and i is is the set of all complex numbers of the form $a + bi$, where a and b are integers. Hence the notation $\mathbb{Z}[i]$ in Example 2.12.2.
5. Show that the subring of $\mathbb{Q}$ generated by $\mathbb{Z}$ and $1/2$ is the set of all rational numbers that have the property that when written in the reduced form a/b with $\gcd(a, b) = 1$, the denominator b is a power of 2. This explains the notation $\mathbb{Z}[1/2]$ in Example 2.12.3.
6. Prove that the subring of $\mathbb{R}$ generated by $\mathbb{Q}[\sqrt{2}]$ and $\sqrt{3}$ is precisely the ring of Example 2.12.4. Thus, this ring should be denoted $\mathbb{Q}[\sqrt{2}][\sqrt{3}]$. We will often avoid using the second pair of brackets and simply refer to this ring as $\mathbb{Q}[\sqrt{2}, \sqrt{3}]$.
7. In Lemma 2.14, suppose a is actually in S. Can you prove that the ring generated by S and a is just S?

Remark 2.16

It is worth emphasizing the difference between a *polynomial expression in a with coefficients in S* and a *polynomial in the variable x with coefficients in S*. The difference is as follows: it is quite possible for two different polynomials expressions in a with coefficients in S such as $s_0 + s_1 a + \cdots + s_n a^n$ and $s'_0 + s'_1 a + \cdots + s'_m a^m$ to be equal without having $n = m$ and $s_0 = s'_0$, $s_1 = s'_1, \ldots, s_n = s'_n$. That is, refering to the integers n and m as "degrees" and the elements s_i and s'_i as "coefficients," *it is possible for two polynomial expressions in a to be equal even though their degrees are not the same or their coefficients are not the same.* However, two polynomials in the variable x with coefficients in S, say $s_0 + s_1 x + \cdots + s_n x^n$ and $s'_0 + s'_1 x + \cdots + s'_m x^m$ are equal only if $n = m$ and $s_1 = s'_1, \ldots, s_n = s'_n$, in other words, only if their degrees are equal and their coefficients are equal.

In particular, given two polynomial expressions either of different degrees or with different coefficients, one cannot automatically conclude that they must be unequal, they may just turn out to be equal!

Here is an example. Taking $S = \mathbb{Q}$ and $a = \sqrt{2}$, consider the following two polynomial expressions in $\sqrt{2}$ with coefficients in $\mathbb{Q}$: $1 + (\sqrt{2}) + (\sqrt{2})^2 + (\sqrt{2})^3$, and $3 + 3(\sqrt{2})$. They are equal! Yet, the first has degree 3 and the second has degreee 1.

The reason why the two expressions are equal, of course, is that $(\sqrt{2})^2$ is the same as 2, so we can substitute 2 wherever we see $(\sqrt{2})^2$ in our expressions. In other words, the reason why it is possible for two polynomial expressions in $\sqrt{2}$ with coefficients in $\mathbb{Q}$ to be equal is that $\sqrt{2}$ satisfies the equation $t^2 - 2 = 0$, that is, the polynomial expression $(\sqrt{2})^2 - 2$ equals zero. By contrast, if x is a variable, a polynomial in x such as $s_0 + s_1x + \cdots + s_nx^n$ can never be zero unless all the coefficients are zero.

Integral Domains and Fields

In passing from the concrete example of the integers to the abstract definition of a ring, observe that we have introduced some phenomena that at first seem pathological. The first, which we have already pointed out explicitly and is already present in $M_2(\mathbb{R})$, is noncommutativity of multiplication. The second, which is also present in $M_2(\mathbb{R})$, and examples of which you have seen as far back as in the Introduction, is the existence of zero-divisors.

Definition 2.17
A *zero-divisor* in a ring R is a nonzero element a for which there exists a nonzero element b such that either $a \cdot b = 0$ or $b \cdot a = 0$.

Just as noncommutativity of multiplication, on closer observation, turns out to be quite a natural phemomenon after all, the existence of zero-divisors is really not very pathological either. It merely *seems* so because most of our experience has been restricted to various rings that appear as subrings of the complex numbers.

Besides matrix rings (try to discover lots of zero-divisors in $M_2(\mathbb{R})$ for yourselves), zero-divisors occur in several rings that arise naturally in mathematics, including many *commutative* ones. For instance, the direct product of two rings always contains zero-divisors (see Example 2.7.10 above). Also, (see Examples 2.7.8 and 2.7.9), $\mathbb{Z}/4\mathbb{Z}$ contains zero-divisors: $[2]_4 \cdot [2]_4 = [0]_4$! In fact, as long as n is not prime, you should be able to discover zero-divisors in any of the rings $\mathbb{Z}/n\mathbb{Z}$. (It can be proved, however, that $\mathbb{Z}/n\mathbb{Z}$ *cannot* have zero-divisors if n is prime.)

On the other hand, there is no doubt that the absence of zero-divisors in a ring indeed makes the ring relatively easy to work with. If, in addition, such a ring is also commutative, it becomes *exceptionally* nice to work with. With this in mind, we make the following definition:

Definition 2.18
An integral domain is a commutative ring with no zero-divisors.

(Convince yourselves that a ring has no zero-divisors if and only if whenever $a \cdot b = 0$ for two elements a and b in the ring, then either a must be 0 or else b must be 0.)

$\mathbb{Z}$, $\mathbb{Q}$, $\mathbb{R}$, and $\mathbb{C}$ are all obvious examples of integral domains. (Again, we are simply invoking our knowledge of these rings when we make this claim.) Is $\mathbb{R}[x]$ an integral domain? More generally, if R is an arbitrary ring, can you determine necessary and sufficient conditions on R that will guarantee that $R[x]$ has no zero-divisors? (See the notes on page 60 for a definition of $R[x]$, and for some discussions that may help you answer this question.)

Notice that any subring S of an integeral domain R must itself be an integral domain. (If $ab = 0$ holds in S for some nonzero elements a and b, then viewing a and b as elements of R, we would find $ab = 0$ in R, which is a contradiction, since R is an integral domain.) In particular, *any subring of* $\mathbb{C}$ *is an integral domain.*

Now suppose S is a subring of R and suppose that S (note!) is an integral domain. Must R also be an integral domain?

Integral domains have one nice property: one can always cancel elements from both sides of an equation! More precisely, we have the following:

Lemma 2.19
Let R be an integral domain, and let a be a nonzero element of R. If $ab = ac$ for two elements b and c in R, then $b = c$.

Proof Write $ab = ac$ as $a(b - c) = 0$. Since $a \neq 0$ and since R is an integral domain, $b - c = 0$, or $b = c$! □

Now, integral domains are definitely very nice rings, but one can go out on a limb and require that rings be even nicer! While considering exactly what we mean by "even nicer," we come to a mathematical object that will occupy a significant portion of our time in this book.

While performing the usual arithmetic operations on the integers, you may have already discovered a crucial property that the integers do not have, and for which it is necessary to go to the rationals (see Example 2.7.1): one cannot divide an arbitrary integer by another arbitrary (nonzero) integer and expect to get an integer as the result. (For instance, there is no integer that represents the result of dividing 3 by 5.) If one were to order rings according to how easy they are to understand and work with, one finds that commutative rings are easier to understand and work with than noncommutative ones, and among commutative rings, one finds that integral domains are easier to understand and work with than those with zero-divisors. Now, among integral domains, the easiest rings to understand and work with are those in which one can also *divide*. Because of the absolute importance of such rings, they are given a special name and singled out for study.

Let us look at the process of dividing two integers a little closer. To divide 3 by 5 is really to multiply together 3 and $1/5$ (just as to subtract, say, 6 from 9 is really to add together 9 and -6). The reason this cannot be done within the context of the integers is that $1/5$ is not an integer. (After all, if $1/5$ were an integer, then the product of 3 and $1/5$ would also be an integer.) Now let us look at $1/5$ a different way. $1/5$ has the property that $1/5 \cdot 5 = 5 \cdot 1/5 = 1$. In other words, $1/5$ *is the inverse of* 5 *with respect to multiplication* (just as -6 is the inverse of 6 with respect to addition). Putting all this together, the reason that we cannot divide within the context of the integers is that given an arbitrary (nonzero integer) m, one cannot find the multiplicative inverse of m, that is, one cannot find an integer n such that $m \cdot n (= n \cdot m) = 1$. With this in mind, we have the following definition:

Definition 2.20

A *field* is an integral domain in which for every nonzero element a, there exists an element b (depending on a) such that $a \cdot b = 1$. The element b is usually denoted either by "$1/a$" or by "a^{-1}."

(For a comment on this definition, see the notes on page 62.)

We will often use the letter F to denote a field. The set of nonzero elements of a field F is often denoted by F^*.

Remark 2.21

If F is a field, can you prove that F^* is a group with respect to multiplication?

Remark 2.22

Notice that 0 can never have a multiplicative inverse, since $a \cdot 0 = 0$ for any a. (See Remark 2.8.) We describe this by saying that *division by 0 is not defined.*

Perhaps the most familiar example of a field is $\mathbb{Q}$. We have already seen that it is a ring (Example 2.7.1) The multiplicative inverse of the nonzero rational number m/n is, of course, n/m.

Examples 2.23

Here are other examples of fields:

1. The reals, $\mathbb{R}$.
2. The complex numbers, $\mathbb{C}$. Can you write the inverse of the nonzero number $a + ib$ as $c + id$ for suitable real numbers c and d?
3. $\mathbb{Q}[\sqrt{2}]$. Show that this ring is actually a field by explicitly exhibiting the multiplicative inverse of $a + b\sqrt{2}$, where a and b are not both zero. (Think in terms of rationalizing denominators.) Is $\mathbb{Z}[\sqrt{2}]$ a field?
4. $\mathbb{Q}[i]$. Why? Is $\mathbb{Z}[i]$ a field?
5. Here is a new example: the set of rational functions with coefficients from the reals, $\mathbb{R}(x)$. (Note the parentheses around the x.) This is the set of all quotients of polynomials with coefficients from the reals, that is, the set $\left\{ \dfrac{f(x)}{g(x)} \right\}$, where $f(x)$ and $g(x)$ are

elements of $\mathbb{R}[x]$, and $g(x) \neq 0$. Addition and multiplication in $\mathbb{R}(x)$ are similar to addition and multiplication in $\mathbb{Q}$—

$$\frac{f_1(x)}{g_1(x)} + \frac{f_2(x)}{g_2(x)} = \frac{f_1(x) \cdot g_2(x) + f_2(x) \cdot g_1(x)}{g_1(x) \cdot g_2(x)},$$

and

$$\frac{f_1(x)}{g_1(x)} \cdot \frac{f_2(x)}{g_2(x)} = \frac{f_1(x) \cdot f_2(x)}{g_1(x) \cdot g_2(x)}.$$

The multiplicative inverse of the nonzero element $\left\{\frac{f(x)}{g(x)}\right\}$ is just $\left\{\frac{g(x)}{f(x)}\right\}$.

Just as we considered subrings of a given ring, we can consider *subfields* of a given field. The idea is very simple: given a field K, first of all view it as a ring. (Remember, every field is first a ring.) Hence, we can talk of subrings of K. We now have the following definition:

Definition 2.24

A subset F of K is called a *subfield of* K if F is a subring of K and is itself a field. If F is a subfield of K, we often call K an *extension field of* F, and refer to the pair (F, K) as the *field extension* K/F.

The difference between being a sub*ring* of K and a sub*field* of K is as follows: Suppose R is a subring of K. Given a nonzero element a in R, its multiplicative inverse $1/a$ certainly exists in K (why?). *However,* $1/a$ *may not live inside* R. If $1/a$ happens to live inside R, we say that a has a multiplicative inverse in R itself. Now, if every nonzero a in R has a multiplicative inverse in R itself, then by Definition 2.20 (why is R an integral domain?), R is a field. Therefore, by Definition 2.24 above, R is then a subfield of K.

Thus, $\mathbb{Q}$ is a subfield of $\mathbb{R}$, but $\mathbb{Z}$ is only a subring of $\mathbb{R}$; it is not a subfield of $\mathbb{R}$. Similarly, $\mathbb{R}$ is a subfield of $\mathbb{C}$. (Is $\mathbb{R}$ a subring of $\mathbb{C}$?) $\mathbb{Q}[\sqrt{2}]$ is a subfield of $\mathbb{C}$. In fact, more is true—$\mathbb{Q}$ is a subfield of $\mathbb{Q}[\sqrt{2}]$, which in turn is a subfield of $\mathbb{R}$, which in turn is a subfield of $\mathbb{C}$. (By contrast, is $\mathbb{Q}[i]$ a subfield of $\mathbb{R}$? Of $\mathbb{C}$? Is $\mathbb{Z}[i]$ a subfield of $\mathbb{R}$? Of $\mathbb{C}$?)

As it turns out, there is a whole host of subfields of the complex numbers, all of which, in turn, contain the rationals as a subfield.

The structure of some of these fields will be crucial to the study of constructibility (as well as to the study of other issues in mathematics). We will develop ideas in the following chapters that will help us understand some of this structure.

Exercises

1. How many different binary operations can be defined on the set $\{0, 1\}$? Now select some of these binary operations and check whether they are associative or commutative. How many binary operations can be constructed on a set T that has n elements?

2. Starting from the ring axioms, prove that the properties stated in Remark 2.8 hold for any ring R.

3. Consider the subset S of $\mathbb{Z}$ consisting of the positive even integers, that is, the set $\{2n | n \in \mathbb{Z} \text{ and } n > 0\}$. Show that S is closed with respect to both addition and multiplication. Does this make S a subring of $\mathbb{Z}$? Next, consider the set T of all nonnegative integers. Show that T is also closed with respect to addition and multiplication. Clearly, T contains 1. Does this make T a subring of $\mathbb{Z}$?

4. If p is any prime, show that the subring of $\mathbb{R}$ generated by $\mathbb{Q}$ and $\sqrt{p}$, that is, $\mathbb{Q}[\sqrt{p}]$, is precisely the set $\{a + b\sqrt{p} | a, b \in \mathbb{Q}\}$. Show that $a + b\sqrt{p} = 0$ if and only if $a = 0$ and $b = 0$. Use this to show that if an element x of $\mathbb{Q}[\sqrt{p}]$ can be expressed as both $a + b\sqrt{p}$ and $a' + b'\sqrt{p}$ for rational numbers a, b, a', and b', then a must equal a' and b must equal b'. (In other words, the expression of x as $a + b\sqrt{p}$ is *unique*.) We know that $\mathbb{Q}[\sqrt{p}]$ is an integral domain. (Remember, no subring of $\mathbb{R}$ can have zero-divisors.) Show that $\mathbb{Q}[\sqrt{p}]$ is actually a field, by explicitly displaying the multiplicative inverse of the nonzero element $a + b\sqrt{p}$.

5. Reformulate Problem 4 for the ring $\mathbb{Q}[\sqrt{-p}]$, and solve it.

6. Consider the ring $M_n(\mathbb{R})$ (for $n \geq 2$). Let $e_{i,j}$ denote the matrix with 1 in the (i, j)-th slot and 0 everywhere else. Study the case of 2×2 matrices and guess at a formula for the product $e_{i,j} \cdot e_{k,l}$.

(You need not try to prove formally that your formula is correct, but after you have made your guess, substitute various values for i, j, k, and l and test your guess.)

7. The following is designed to show that if a, b, c, and d are rational numbers, then $a + b\sqrt{2} + c\sqrt{3} + d\sqrt{6} = 0$ if and only if a, b, c, and d are all zero.

   (a) Show that $\sqrt{3/2}$ is not rational. (This is similar to showing that $\sqrt{p}$ is not rational for any prime p.

   (b) Show that $\sqrt{3} \notin \mathbb{Q}[\sqrt{2}]$. (Hint: Assume that $\sqrt{3} \in \mathbb{Q}[\sqrt{2}]$. Then there must exist rational numbers x and y such that $\sqrt{3} = x + y\sqrt{2}$. Square both sides and arrive at a contradiction. You will need to invoke a fact about $\mathbb{Q}[\sqrt{2}]$ that you were asked to prove in Example 2.7.3, as well as the results of Chapter 1, Exercise 17, and part 7a above.)

   (c) Now assume that $a + b\sqrt{2} + c\sqrt{3} + d\sqrt{6} = 0$ for some choice of rational numbers a, b, c, and d. Write this as $(a + b\sqrt{2}) + \sqrt{3}(c + d\sqrt{2}) = 0$. Prove that $c + d\sqrt{2}$ must be zero. (Hint: Argue that otherwise we can write $\sqrt{3} = -\dfrac{a + b\sqrt{2}}{c + d\sqrt{2}}$. Why is this last equality a contradiction?)

   (d) Conclude that this forces $a = b = c = d = 0$.

   (e) Observe that if $a = b = c = d = 0$ then $a + b\sqrt{2} + c\sqrt{3} + d\sqrt{6} = 0$ trivially. This proves the assertion stated at the beginning.

8. We will prove in this exercise that $\mathbb{Q}[\sqrt{2}, \sqrt{3}]$ (see Example 2.15.6 for this notation) is actually a field.

   (a) You know that if a and b are rational numbers, then $(a + b\sqrt{2}) \cdot (a - b\sqrt{2})$ is also rational. (Why?) Similarly, if c and d are rational numbers, then $(c + d\sqrt{3}) \cdot (c - d\sqrt{3})$ is also rational. Now show the following: if a, b, c, and d are all rational numbers, then the product of the four terms

$$\begin{array}{ccc} (a + b\sqrt{2} + c\sqrt{3} + d\sqrt{6}) & \cdot & (a + b\sqrt{2} - c\sqrt{3} - d\sqrt{6}) \quad \cdot \\ (a - b\sqrt{2} + c\sqrt{3} - d\sqrt{6}) & \cdot & (a - b\sqrt{2} - c\sqrt{3} + d\sqrt{6}) \end{array}$$

   is also rational. (This just involves multiplying out all the terms above—do it! However, you can save yourselves a lot of work by multiplying the first two terms together, using the formula $(x + y)(x - y) = x^2 - y^2$, and then multiplying the remaining two terms together, and looking out for patterns.)

(b) Now show using part 8a above that $\mathbb{Q}[\sqrt{2}, \sqrt{3}]$ is a field. (Hint: Given a nonzero element $a + b\sqrt{2} + c\sqrt{3} + d\sqrt{6}$ in $\mathbb{Q}[\sqrt{2}, \sqrt{3}]$, first note that by Exercise 7 above, none of $(a + b\sqrt{2} - c\sqrt{3} - d\sqrt{6})$, $(a - b\sqrt{2} + c\sqrt{3} - d\sqrt{6})$ or $(a - b\sqrt{2} - c\sqrt{3} + d\sqrt{6})$ can be zero—why? Now, in the case of $\mathbb{Q}[\sqrt{2}]$, one finds the inverse of $x + y\sqrt{2}$ by multiplying both the numerator and the denominator of the fraction $\frac{1}{x+y\sqrt{2}}$ by $x - y\sqrt{2}$ and taking advantage of the fact that $(x + y\sqrt{2})(x - y\sqrt{2})$ is rational. What ideas do you get from part 8a above?)

9. If R is an *arbitrary* ring, a nonzero element a is said to be *invertible* or to have a *multiplicative inverse* if there exists an element $b \in R$ such that $ab = ba = 1$. (Notice that for an arbitrary ring, it is not enough to insist that $ab = 1$, we also need ba to equal 1.) Prove that if an element a in an arbitrary ring R is a zero-divisor, then a cannot be invertible. Give an example to show that the converse is false.

10. Here is an example of a ring in which elements do not factor uniquely into a product of primes! Consider the subring of $\mathbb{C}$ generated by $\mathbb{Z}$ and $\sqrt{-5}$, namely, $\mathbb{Z}[\sqrt{-5}]$. By arguments nearly identical to those that you must have used in Exercise 5 above, every element in this ring can be written *uniquely* as $a + b\sqrt{-5}$ for suitable integers a and b. We define a function N from $\mathbb{Z}[\sqrt{-5}]$ to $\mathbb{Z}$ as follows: $N(a + b\sqrt{-5}) = a^2 + 5b^2$. (Notice that $a^2 + 5b^2$ is just $(a + b\sqrt{-5}) \cdot (a - b\sqrt{-5})$.)

(a) Show that N is *multiplicative*, that is, $N(xy) = N(x)N(y)$ for any two elements x and y of $\mathbb{Z}[\sqrt{-5}]$.

(b) Show that if x in $\mathbb{Z}[\sqrt{-5}]$ is such that $N(x) = 1$, then x must be ± 1.

(c) If R is any commutative ring, a *unit* of R is defined as any element x of R for which there exists a y in R such that $xy = 1$. Thus, in the terminology of Exercise 9 above, a unit in a commutative ring is just an invertible element. (As examples of units, note that the units of $\mathbb{Z}$ are precisely ± 1, whereas in

$\mathbb{Q}$ or any other field, every nonzero element is a unit.) Use part 10a to show that if x is a unit in $\mathbb{Z}[\sqrt{-5}]$, then $N(x)$ must be 1.

(d) Use parts 10b and 10c above to show that if x is a unit in $\mathbb{Z}[\sqrt{-5}]$, then x can only be ± 1.

(e) If R is a commutative ring, an *irreducible* in R is a nonzero element x such that if $x = bc$ for two elements b and c, then either b or c must be a unit. (It turns out that this is the correct generalization of the concept of primes that is needed to study unique factorization in arbitrary rings.) Also, just as in $\mathbb{Z}$, we say an element b in an arbitrary commutative ring R *divides* an element a (or is a *divisor* of a) if there exists an element c in R such that $a = bc$. Using part 10d, show that if x is an irreducible element in $\mathbb{Z}[\sqrt{-5}]$, then the only divisors of x are ± 1 and $\pm x$. (Thus, at least in $\mathbb{Z}[\sqrt{-5}]$, it is clear that irreducible elements are just like primes.)

(f) Show that if x in $\mathbb{Z}[\sqrt{-5}]$ is such that $N(x)$ is a prime integer, then x must be irreducible.

(g) Show that there is no element x in $\mathbb{Z}[\sqrt{-5}]$ with $N(x) = 2$. Similarly, show that there is no element x with $N(x) = 3$.

(h) Show that 2 is irreducible in $\mathbb{Z}[\sqrt{-5}]$. (Hint: Suppose $2 = xy$. Then $4 = N(2) = N(x)N(y)$, as N is multiplicative. Study the various factorizations of 4 and use the previous parts.)

(i) Similarly, show that 3 is irreducible in $\mathbb{Z}[\sqrt{-5}]$.

(j) Study the various factors of $N(1 + \sqrt{-5})$ and of $N(1 - \sqrt{-5})$ and show that both $1 + \sqrt{-5}$ and $1 - \sqrt{-5}$ are irreducible.

(k) Two irreducibles x and y in a commutative ring R are said to be *associates* if $x = yu$ for some unit u. Part 10d shows that in the ring $\mathbb{Z}[\sqrt{-5}]$, two elements x and y are associates if and only if $x = \pm y$. Now use the fact that every element in $\mathbb{Z}[\sqrt{-5}]$ is uniquely expressible as $a + b\sqrt{-5}$ to show that neither 2 nor 3 is an associate of either $1 + \sqrt{-5}$ or $1 - \sqrt{-5}$.

(l) A commutative ring R is said to possess unique prime factorization if every element $a \in R$ that is not a unit factors into a product of irreducibles, and if $a = x_1 x_2 \cdots x_s$ and $a = y_1 y_2 \cdots y_t$ are two factorizations of a into irreducibles, then s must equal t, and after relabeling if necessary, each x_i must be an associate of the corresponding y_i. (Again, it turns out that this is the correct generalization of the concept of unique prime factorization in the integers to arbitrary commutative rings.) Prove that $\mathbb{Z}[\sqrt{-5}]$ does not possess unique prime factorization by considering two different factorizations of 6 into irreducibles. (Hint: Look at parts 10h, 10i, 10j, and 10k.)

Notes

Remarks on Example 2.7.1 Every nonzero element in $\mathbb{Q}$ has a multiplicative inverse, that is, given any $q \in \mathbb{Q}$ with $q \neq 0$, we can find a rational number q' such that $qq' = 1$. The same cannot be said for the integers: not every nonzero integer has a multiplicative inverse within the integers. For example, there is no integer a such that $2a = 1$, so 2 does not have a multiplicative inverse.

Remarks on Example 2.7.3 The sum and product of any two elements $a + b\sqrt{2}$ and $c + d\sqrt{2}$ of $\mathbb{Q}[\sqrt{2}]$ are (respectively) $(a + c) + (b + d)\sqrt{2}$ and $(ac + 2bd) + (ad + bc)\sqrt{2}$. Since $a + c$, $b + d$, $ac + 2bd$ and $ad + bc$ are all rational numbers, the sum and product also lie in $\mathbb{Q}[\sqrt{2}]$. Thus, the standard method of adding and multiplying two real numbers of the form $x + y\sqrt{2}$ with x and y in $\mathbb{Q}$ indeed gives us binary operations on $\mathbb{Q}[\sqrt{2}]$. (In the language of the next section, $\mathbb{Q}[\sqrt{2}]$ is closed under addition and multiplication.) Now suppose you were trying to prove that, say, addition in $\mathbb{Q}[\sqrt{2}]$ is associative, that is, for any u, v, and w in $\mathbb{Q}[\sqrt{2}]$, $(u + v) + w = u + (v + w)$. Notice that in addition to being in $\mathbb{Q}[\sqrt{2}]$, u, v, and w are also *real numbers*. Since associativity holds in the reals, we find upon viewing u, v, and w as real numbers that $(u + v) + w = u + (v + w)$. Now viewing u, v, and w in this equation back again as elements of $\mathbb{Q}[\sqrt{2}]$, we find that associativity holds in $\mathbb{Q}[\sqrt{2}]$! This same argument holds for associativity of multiplication and distributivity of multiplication over addition. To prove that $a + b\sqrt{2} = 0$ iff $a = 0$ and $b = 0$, proceed as follows: If b is not zero, $a + b\sqrt{2} = 0$ yields $\sqrt{2} = -a/b$. Since $-a/b$ is a rational number, this contradicts Chapter 1, Exercise 17, so b must be zero. But if $b = 0$, $a + b\sqrt{2} = 0$ reads $a = 0$, so we find that both a and b are zero.

Remarks on Example 2.7.4 As in the previous example, if b is not zero, we can write $\sqrt{-1} = a/b$, and squaring both sides, we find $-1 = a^2/b^2$. The right hand side is positive, since both a^2 and b^2 are positive (they are squares). But the left hand side is negative. Because of this contradiction, b must be zero. As before, we find that a is also zero.

Remarks on Examples 2.7.5 Given two elements a and b in $\mathbb{Z}_{(2)}$, write a as x/y where $\gcd(x,y) = 1$ and y is odd. (Why can you do this?) Write b as u/v where $\gcd(u,v) = 1$ and v is odd. Then $a+b = (xv+yu)/yv$. This fraction may not be reduced, but notice that yv, being a product of two odd integers, is odd. After you cancel all common factors from $(xv + yu)$ and yv, the resultant fraction will still have an odd denominator (why?). Hence $a + b$ will be in $\mathbb{Z}_{(2)}$. In a similar way, show that ab (gotten by the usual multiplication of two rational numbers) will also be in $\mathbb{Z}_{(2)}$. Now that you have two binary operations on $\mathbb{Z}_{(2)}$, you can check that the ring axioms hold. As with previous examples, associativity and distributivity follow from the fact that they hold for the rationals. Notice that the fact that the product of two odd integers is odd was essential in showing that both $a + b$ and ab lie in $\mathbb{Z}_{(2)}$. How could we generalize this? Rewrite this property in the contrapositive form, yv is even implies either y or v is even, that is, $2|yv$ implies $2|y$ or $2|v$. If we could find another integer n that has the property that $n|yv$ implies $n|y$ or $n|v$, we could use the same arguments to show that $\mathbb{Z}_{(n)}$ is also a ring. (Assuming you found such an integer n, how would you define $\mathbb{Z}_{(n)}$?) Can you think of other integers that have this property? (Hint: You have come across such integers in the previous chapter.)

Remarks on Example 2.7.6 For $n = 1$, $M_n(\mathbb{R})$ is just $\mathbb{R}$, so it is commutative. For all other n, $M_n(\mathbb{R})$ is noncommutative. Given A in $M_n(\mathbb{R})$, write A as $((a_{i,j}))$. (Recall what this notation means: you are referring to the (i,j)-th entry as "$a_{i,j}$.") Similarly, write B as $((b_{i,j}))$ and C as $((c_{i,j}))$. Consider $(A + B) + C$. What is the (i,j)-th entry of this resultant matrix? It is $(a_{i,j} + b_{i,j}) + c_{i,j}$. On the other hand, what is the (i,j)-th entry of $A + (B + C)$? It is $a_{i,j} + (b_{i,j} + c_{i,j})$. Are the two (i,j)-th entries equal on both sides? Yes! Why? Because $a_{i,j}$, $b_{i,j}$, and $c_{i,j}$ are just real numbers, and *since addition is associative in* $\mathbb{R}$, $(a_{i,j} + b_{i,j}) + c_{i,j} = a_{i,j} + (b_{i,j} + c_{i,j})$! Since this is true for every pair (i,j), we find that $(A + B) + C = A + (B + C)$. (Notice how the associativity of addition in $M_n(\mathbb{R})$ *depends* on the associativity of addition in $\mathbb{R}$.) In a similar manner, try to prove the distributive property of multiplication over addition for $M_n(\mathbb{R})$; your proof should invoke the fact that distributivity holds in $\mathbb{R}$. Actually, if R is any ring, $M_n(R)$ is also a ring. It is noncommutative if $n \geq 2$. When $n = 1$, $M_n(R)$ is just R, so for $n = 1$, $M_n(R)$ is commutative if and only if R is commutative.

Remarks on Example 2.7.7 For any ring R, we can consider the set of polynomials with coefficients in R with the usual definition of addition and multiplication of polynomials. This will be a ring, with additive identity the constant polynomial 0 and multiplicative identity the constant polynomial 1. If R is commutative, $R[x]$ will also be commutative. (Why? Play with two general polynomials $f = \sum_{i=0}^{n} f_i x^i$ and $g = \sum_{j=0}^{m} g_j x^j$ and study fg and gf.) If R is not commutative, $R[x]$ will also not be commutative. To see this last assertion, suppose a and b in R are such that $ab \neq ba$. Then viewing a and b as constant polynomials in $R[x]$, we find that we get two different products of the "polynomials" a and b depending on the order in which we multiply them!

Here is something strange that can happen with polynomials with coefficients in an arbitrary ring R. First, the *degree* and *highest coefficient* of polynomials in $R[x]$ (where R is arbitrary) are defined exactly as for polynomials with coefficients in the reals. Now over $\mathbb{R}[x]$, if $f(x)$ and $g(x)$ are two nonzero polynomials, then $deg(f(x)g(x)) = deg(f(x)) + deg(g(x))$. But for an arbitrary ring R, the degree of $f(x)g(x)$ *can be less than* $deg(f(x)) + deg(g(x))$!

To see why this is, suppose $f(x) = f_n x^n$ + lower-degree terms (with $f_n \neq 0$), and suppose $g(x) = g_m x^m$ + lower-degree terms (with $g_m \neq 0$). On multiplying out $f(x)$ and $g(x)$, the highest power of x that will show up in the product is x^{n+m}, and its coefficient will be $f_n g_m$. If we are working in $\mathbb{R}$, then $f_n \neq 0$ and $g_m \neq 0$ will force $f_n g_m$ to be nonzero, so the degree of $f(x)g(x)$ will be exactly $n + m$. But over arbitrary rings, it is quite possible for $f_n g_m$ to be zero even though f_n and g_m are themselves nonzero. (You have already seen examples of this in matrix rings. Elements a and b in a ring R such that $a \neq 0$ and $b \neq 0$ but $ab = 0$ will be referred to later in the chapter as zero-divisors.) When this happens, the highest *nonzero* term in $f(x)g(x)$ will be something lower than the x^{n+m} term, so the degree of $f(x)g(x)$ will be less than $n + m$!

Clearly, this phenomenon will not occur if the coefficient ring R does not have any zero-divisors. As will be explained further along in the chapter, *fields* do not have any zero-divisors (they are special cases of *integral domains.*) Hence if F is a field and $f(x)$ and $g(x)$ are two nonzero polynomials in $F[x]$, then $deg(f(x)g(x)) = deg(f(x)) + deg(g(x))$.

Remarks on Example 2.7.10 The additive identity is $(0, 0)$ and the multiplicative identity is $(1, 1)$. What is the product of $(1, 0)$ and $(0, 1)$? Of $(2, 0)$ and $(0, 2)$?

Remarks on Definition 2.10 The requirement that 1 be in S arises from a rather nasty technical point that can be ignored during a first reading. If you are curious, recall first that "1" is merely notation for the multiplicative identity of R; we could just as easily have referred to it as "e" or something else all along. It turns out that if we defined subrings without the condition that 1 be in S, then it is possible for S to be a subring of R (under this hypothetical definition) with S and R having *different* multiplicative identities! This is a scenario we wish to avoid, and it turns out that insisting that the multiplicative identity of R (namely 1) be in S will take care of this problem. At the same time, it turns out that no such precaution needs to be taken for the *additive* identity—it can be proved that the additive identities of R and S will necessarily be equal. This is of course all too pedantic for a first go around—we would do best by just accepting the definition above and getting on with our lives!

Remarks on Examples 2.12.3 Since every integer a can be written as $a/1$, and since 1 of course is 2^0, $\mathbb{Z} \subseteq \mathbb{Z}[1/2]$. Since 2 does not divide 1, every integer a ($= a/1$) is also in $\mathbb{Z}_{(2)}$. Hence, $\mathbb{Z}[1/2] \cap \mathbb{Z}_{(2)}$ certainly contains $\mathbb{Z}$. Now let x be any rational number in $\mathbb{Z}[1/2] \cap \mathbb{Z}_{(2)}$. Since $x \in \mathbb{Z}[1/2]$, x can be written in the reduced form $a/2^n$, for some integer a and some $n \geq 0$. If $n > 0$, then x cannot be in $\mathbb{Z}_{(2)}$. Hence $n = 0$, that is, $x \in \mathbb{Z}$. It follows that $\mathbb{Z}[1/2] \cap \mathbb{Z}_{(2)}$ is precisely $\mathbb{Z}$.

Remarks on Example 2.12.8 Denote the function that sends r to $\text{diag}(r)$ by f. Then f is bijective. Also, $f(r + s) = f(r) + f(s)$, and $f(rs) = f(r)f(s)$. Moreover, $f(0)$ is just the zero matrix, and $f(1)$ is just the identity matrix. What all this means is that the subring of matrices of the form $\text{diag}(r)$ is *essentially identical* to $\mathbb{R}$ in the following sense: to each real number r we can associate the matrix $\text{diag}(r)$ and vice versa, and under this correspondence, the rule for adding and multiplying two real numbers r and s is the same as the rule for adding and multiplying the two matrices diag(r) and diag(s). In more technical language, the two rings $\mathbb{R}$ and the set of matrices of the form diag(r) are said to be *isomorphic*.

Remarks on the proof of Lemma 2.14 How would you multiply two polynomial expressions in a with coefficients in S? Suppose $p_1 = s_0 + s_1a + s_2a^2 + \cdots + s_n$ is one such expression, and $p_2 = t_0 + t_1a + t_2a^2 + \cdots + t_ma^m$ is another such expression. Their product is $(s_0t_0) + (s_0t_1 + s_1t_0)a + (s_0t_2 + s_1t_1 + s_2t_0)a^2 + \cdots + (s_nt_m)a^{n+m}$. Since s_0t_0, $s_0t_1 + s_1t_0$, $s_0t_2 + s_1t_1 + s_2t_0$, etc. are all in S, the product is also a polynomial expression in a with coefficients in S. The proof that $p_1 + p_2$ is a polynomial expression is even simpler!

Remarks on Definition 2.20 Most textbooks define a field to be a commutative ring in which for every nonzero a, there is a nonzero element b for which $ab = 1$. In other words, the extra condition that we have imposed, namely that the ring in question first be an integral domain, is omitted by most textbooks. This is because this extra condition is not really required—one can show using Exercise 9 that any commutative ring in which every nonzero element a is invertible (see Exercise 9 for the meaning of this term) must *necessarily* be an integral domain. The reason we have chosen to define a field as an *integral domain* in which every nonzero element is invertible is to highlight the hierarchical nature of the objects that we have been considering: rings are fairly general objects, commutative rings are special rings that are nicer to deal with, integral domains are special commutative rings that are even nicer, and finally, fields are special integral domains that are nicest of all!

3 CHAPTER Vector Spaces

The whole theory of constructibility depends on the analysis of various field extensions K of $\mathbb{Q}$. Now, given such a field extension, one of the first things one would like to do is somehow measure *how big* K is relative to $\mathbb{Q}$. Of course, "how big" is a loose term, and we need to make this question more precise. To do so, we need to invoke some ideas from the theory of vector spaces.

Recall from elementary linear algebra the notation $\mathbb{R}^2$ for 2-dimensional xy space and $\mathbb{R}^3$ for 3-dimensional xyz space. A *vector* in $\mathbb{R}^2$ (respectively $\mathbb{R}^3$) is an arrow with its base at the origin and its tip at some point in $\mathbb{R}^2$ (respectively $\mathbb{R}^3$). If v and w are vectors, then we *add* v and w using the parallelogram law. We know that this process of addition is commutative, that is, $v + w = w + v$ for all vectors v and w. Vector addition is also associative, that is, $v + (w + u) = (v + w) + u$ for all vectors v, w, and u. The vector whose base *and* tip are at the origin is denoted 0 (suggestively), and satisfies $v + 0 = 0 + v$ for all vectors v. Finally, for every vector v, the vector we get by inverting v about the origin is denoted $-v$ (also suggestively), and satisfies $v + (-v) = (-v) + v = 0$.

Focusing just on $\mathbb{R}^2$ for convenience, let us stop thinking of $\mathbb{R}^2$ as a geometric object. Instead, since every point of $\mathbb{R}^2$ corresponds to a vector whose tip is at the given point, let us consider $\mathbb{R}^2$ as a set consisting of abstract objects called *vectors*. This set has a binary operation defined on it—addition, where $v + w$ is defined as the vector we get by temporarily reverting to the geometric interpretation of $\mathbb{R}^2$ as a plane and considering the vector obtained as the diagonal of the parallelogram formed by v and w. What do you notice about this set of vectors with this binary operation? The binary operation satisfies all the axioms for an abelian group! Thus, in addition to being a geometric object (the plane), $\mathbb{R}^2$, when considered as a set with a binary operation, has an algebraic structure—it is an abelian group!

But there is more. Let us go back to the interpretation of $\mathbb{R}^2$ as 2-dimensional xy space, and let us recall the notion of scalar multiplication. A *scalar* is any real number, and given a scalar r and a vector v, we multiply r and v according to the following definition—if $r \geq 0$, then $r \cdot v$ is the vector in the same direction as v but whose length is r times the length of v, and if $r < 0$, then $r \cdot v$ is the vector in the opposite direction as v but whose length is $|r|$ times the length of v. What are the properties of scalar multiplication? If r and s are any two scalars, and if v and w are any two vectors, we have the following: $r \cdot (v + w) = r \cdot v + r \cdot w$, $(r + s) \cdot v = r \cdot v + s \cdot v$, $(rs) \cdot v = r \cdot (s \cdot v)$, and $1 \cdot v = v$.

Observe that the set of scalars, namely the real numbers, is a *field*. Now, let us attempt to generalize all this. In the case of $\mathbb{R}^2$ above, we have seen that the geometric interpretation of $\mathbb{R}^2$ as 2-dimensional xy space furnishes us with the notion of vector addition and scalar multiplication, but once these definitions have been furnished, $\mathbb{R}^2$ seems to have an *algebraic* life of its own. For instance, $(\mathbb{R}^2, +)$ is an abelian group, while scalar multiplication has the (algebraic) properties listed above. Could similar sets of objects called *vectors* and *scalars* not arise in different circumstances, with the same properties as the ones listed above, but with the vector addition and scalar multiplication perhaps defined by some process other than a geometric one? The answer is *yes*, and in fact, they arise in vastly different situations. As with the other concepts that we have seen (groups, rings, fields, etc.), it is worth isolating this phenomenon and studying it in its own right.

Definition 3.1

Let F be a field. A *vector space over* F (also called an F*-vector space*) is an abelian group V together with a function $F \times V \to V$ called *scalar multiplication* and denoted $\cdot$ such that for all r and s in F and v and w in V,

1. $r \cdot (v + w) = r \cdot v + r \cdot w$,
2. $(r + s) \cdot v = r \cdot v + s \cdot v$,
3. $(rs) \cdot v = r \cdot (s \cdot v)$, and
4. $1 \cdot v = v$.

The elements of V are called *vectors* and the elements of F are called *scalars*.

Thus, $\mathbb{R}^2$ and $\mathbb{R}^3$ are both vector spaces over $\mathbb{R}$. But what does this formulation of the abstract concept of a vector space really buy us? We will see the answer in the context of field extensions—plenty! But first, let us look at several examples of vector spaces that arise from nongeometric considerations:

Examples 3.2

1. We have looked at $\mathbb{R}^2$ and $\mathbb{R}^3$, why not generalize these, and consider $\mathbb{R}^4$, $\mathbb{R}^5$, etc.? These would of course correspond to higher-dimensional "worlds." It is certainly hard to visualize such spaces, but there is no problem considering them in a purely algebraic manner. Recall that every vector in $\mathbb{R}^2$ can be described *uniquely* by the *pair* (a, b), consisting of the x and y components of the vector. ("Uniquely" means that the vector (a, b) equals the vector (a', b') if and only if $a = a'$ and $b = b'$.) Similarly, every vector in $\mathbb{R}^3$ can be described uniquely by the *triple* (a, b, c), consisting of the x, y, and z components of the vector. Thus, $\mathbb{R}^2$ and $\mathbb{R}^3$ can be described respectively as the set of all pairs (a, b) and the set of all triples (a, b, c), where a, b, and c are arbitrary real numbers. Proceeding analogously, for any positive integer n, we will let $\mathbb{R}^n$ denote the set of *n-tuples* $(a_1, a_2, \ldots, a_n)$, where the a_i are arbitrary real numbers. (As with $\mathbb{R}^2$ and $\mathbb{R}^3$, the understanding here is that two n-tuples $(a_1, a_2, \ldots, a_n)$ and $(a_1', a_2', \ldots, a_n')$ are equal if and only if their respective components are equal, that is, $a_1 = a_1'$, $a_2 = a_2'$, ..., and $a_n = a_n'$.) These n-tuples will be our vectors; how should we add them? Recall that in $\mathbb{R}^2$ we add the vectors (a, b) and (a', b') by adding a and a' together and b and b' together, that is, by adding *componentwise*. (If you are only familiar with the parallelogram law for adding vectors, see Exercise 1.) We will do the same with $\mathbb{R}^n$—we will decree that

$(a_1, a_2, \ldots, a_n) + (a'_1, a'_2, \ldots, a'_n) = (a_1 + a'_1, a_2 + a'_2, \ldots, a_n + a'_n)$. Check that with this definition of addition, $(\mathbb{R}^n, +)$ is an abelian group. What should our scalars be? Just as in $\mathbb{R}^2$ and $\mathbb{R}^3$, let us take our scalars to be the field $\mathbb{R}$. How about scalar multiplication? In $\mathbb{R}^2$, the product of the scalar r and the vector (a, b) is (ra, rb), that is, we multiply each component of the vector (a, b) by the real number r. (Is that so? Check!) We will multiply scalars and vectors in $\mathbb{R}^n$ in the same way: we will decree that the product of the real number r and the n-tuple $(a_1, a_2, \ldots, a_n)$ is $(ra_1, ra_2, \ldots, ra_n)$. Check that this definition satisfies the axioms of scalar multiplication in Definition 3.1. Thus, $\mathbb{R}^n$ is a vector space over $\mathbb{R}$.

2. Now, why restrict the examples above to n-tuples of $\mathbb{R}$? For *any* field F, let F^n stand for the set of n-tuples $(a_1, a_2, \ldots, a_n)$, where the a_i are arbitrary elements of F. Add two such n-tuples componentwise, that is, define addition via the rule $(a_1, a_2, \ldots, a_n) + (a'_1, a'_2, \ldots, a'_n) = (a_1 + a'_1, a_2 + a'_2, \ldots, a_n + a'_n)$. Take the field F to be the field of scalars, and define scalar multiplication just as in $\mathbb{R}^n$: given an arbitrary $f \in F$, and an arbitrary n-tuple $(a_1, a_2, \ldots, a_n)$, define their scalar product to be the n-tuple $(fa_1, fa_2, \ldots, fa_n)$. Check that these definitions of vector addition and scalar multiplication make F^n a vector space over F. Taking $F = \mathbb{C}$ and $n = 2$ for instance, we get *complex 2-space,* which is a natural arena in which to study plane curves.
3. Similarly, for any field F, let F^∞ denote the set of all *infinite-tuples* $(a_0, a_1, a_2, \ldots)$, where the a_i are in F. (It is convenient in certain applications to index the components from 0 rather than 1, but if this bothers you, it is harmless to think of the tuples as $(a_1, a_2, a_3, \ldots)$.) Addition and scalar multiplication are defined just as in F^n, except that we now have infinitely many components. With these definitions, F^∞ becomes an F–vector space.
4. Consider the ring $M_n(\mathbb{R})$. Focusing just on the addition operation on $M_n(\mathbb{R})$, recall that $(M_n(\mathbb{R}), +)$ is an abelian group. (Remember, for any ring R, $(R, +)$ is always an abelian group.) We will treat the reals as scalars. Given any real number r and any matrix $((a_{i,j}))$ in $M_n(\mathbb{R})$, we will define their product to be the matrix $((ra_{i,j}))$. (See the notes on page 91 for a comment on this product.) Verify

that with this definition, $M_n(\mathbb{R})$ is a vector space over $\mathbb{R}$. In a similar manner, if F is any field, $M_n(F)$ will be a vector space over F.

5. Consider the field $\mathbb{Q}[\sqrt{2}]$. Then $(\mathbb{Q}[\sqrt{2}], +)$ is an abelian group (why?). Think of the rationals as scalars. There is a very natural way of multiplying a rational number q with an element $a + b\sqrt{2}$ of $\mathbb{Q}[\sqrt{2}]$, namely, $q \cdot (a + b\sqrt{2}) = qa + qb\sqrt{2}$. With this definition of scalar multiplication, check that $\mathbb{Q}[\sqrt{2}]$ becomes a vector space over the rationals.

   If you probe this example a little harder, you may come up with an apparent anomaly. What exactly is the role of the rationals here? True, we want to think of the rationals as scalars. However, $\mathbb{Q} \subseteq \mathbb{Q}[\sqrt{2}]$, so every rational number is also an element of $\mathbb{Q}[\sqrt{2}]$, and is therefore also a vector! How do we resolve this conflict? As it turns out, there really is nothing to resolve, we merely accept the fact that the rationals have a dual role in this example! When we see a rational number "q" by itself, we want to think of it as $q + 0\sqrt{2}$, that is, we want to think of q as an element of $\mathbb{Q}[\sqrt{2}]$, or in other words, we want to think of q as a vector. However, when we see q in an expression like $q(a + b\sqrt{2})$, we want to think of q as a scalar, that is, something we multiply vectors by!

6. Let us generalize Example 5. What we needed above were that
   (a) $\mathbb{Q}[\sqrt{2}]$ is a field, so $(\mathbb{Q}[\sqrt{2}], +)$ is automatically an abelian group, and
   (b) $\mathbb{Q} \subseteq \mathbb{Q}[\sqrt{2}]$, so that we could use the natural multiplication inside $\mathbb{Q}[\sqrt{2}]$ to multiply any $q \in \mathbb{Q}$ with any $a + b\sqrt{2} \in \mathbb{Q}[\sqrt{2}]$.

   These two facts together gave us a $\mathbb{Q}$–vector space structure on $\mathbb{Q}[\sqrt{2}]$. Now let K/F be *any* field extension. Since K is a field, $(K, +)$ is an abelian group. Next, let us consider multiplication. Given any two elements k and l of K, we know we can multiply the two elements together. However, let us ignore this fact temporarily, and just consider the fact that given any element f of F and any element k of K, we can multiply f and k. (Notice that we have restricted the first element to be from F. However, we have placed no restriction on the second element, it can be any element of K. This is just like considering the multiplication of any $q \in \mathbb{Q}$ and any $a + b\sqrt{2} \in \mathbb{Q}[\sqrt{2}]$ in Example 5 above.) Now note the following properties of this (restricted) multiplication, which are just consequences of the properties of the (unrestricted) multiplication in K: If f and g are any two elements of F, and k and l are any two elements of K, then 1) $f \cdot (k + l) = f \cdot k + f \cdot l$, 2) $(f + g) \cdot k = f \cdot k + g \cdot k$, 3) $(fg) \cdot k = f \cdot (g \cdot k)$, and 4) $1 \cdot k = k$. (In this last property, we consider 1 as an element of F.) What do we notice? If we take the field F as our scalars, $(K, +)$ as our vectors, and the multiplication operation between elements of F and elements of K (that arises from the multiplication operation on K) as scalar multiplication, then, just as in Example 5 above, K becomes an F-vector space!

Also, exactly as in Example 5 above, the elements of F have a dual role, both as scalars and as vectors. When we see an element $f \in F$ by itself, f is playing the role of a vector. But when we see an element $f \in F$ in an expression like $f \cdot k$, f is playing the role of a scalar that is multiplying the vector k!

In particular, taking $F = \mathbb{Q}$, we find that any extension field K of $\mathbb{Q}$ is automatically a vector space over $\mathbb{Q}$. This fact will be central to our study of constructibility.

7. Now let us generalize Example 5 even further, by once again considering the two conditions at the beginning of Example 6. Do we really need the full force of the fact that $\mathbb{Q}[\sqrt{2}]$ is a *field*? No, all we need is the fact that $\mathbb{Q}[\sqrt{2}]$ is a *ring* that contains the field $\mathbb{Q}$; this is enough to provide an abelian group structure on $(\mathbb{Q}[\sqrt{2}], +)$ and to furnish a scalar product between elements of $\mathbb{Q}$ and elements of $\mathbb{Q}[\sqrt{2}]$. Now let R be any ring that contains a field F. Then just as in Example 6 above, $(R, +)$ is an abelian group, and we can use the multiplication in R to define the scalar product between any element f of F and any element r of R. This multiplication clearly satisfies the scalar product axioms in Definition 3.1, so R becomes an F-vector space. Just as in Example 6 above, the elements of F have a dual role, both as scalars and as vectors.

   Here is a familiar instance of this phenomenon. Consider the polynomial ring $\mathbb{R}[x]$. This ring contains $\mathbb{R}$ (since every real number r lives inside $\mathbb{R}[x]$ as the constant polynomial $r + 0x + 0x^2 + \cdots$). Thus, $\mathbb{R}[x]$ is a vector space over $\mathbb{R}$. Explicitly, the scalar product of any real number r and any polynomial $f =$

$\sum_{i=0}^{n} a_i x^i$ (where the a_i are real numbers and n is some nonnegative integer) is the polynomial $\sum_{i=0}^{n} ra_i x^i$. The real numbers have a dual role here: when we see a real number r by itself, we want to think of it as a vector, and when we see it in an expression $r \cdot f$, we want to think of it as a scalar multiplying the vector f.

In the same vein, $F[x]$ is an F–vector space for any field F.

8. Here is an example related to $F[x]$. For any field F and any nonnegative integer n, write $F_n[x]$ for the set of all polynomials in x with coefficients in F whose degrees are *at most* n. Then $F_n[x]$ is an F–vector space. Why?
9. Here is another instance of the situation described in Example 7 above. Take R to be $\mathbb{Q}[\pi]$, the subring of $\mathbb{R}$ generated by $\mathbb{Q}$ and π. (We will see later in Exercise 6 of Chapter 4 that $\mathbb{Q}[\pi]$ is *not* a field; it is only an integral domain.) $\mathbb{Q}[\pi]$ contains $\mathbb{Q}$ (does it?), so $\mathbb{Q}[\pi]$ is a $\mathbb{Q}$–vector space.
10. Now think about this: Suppose V is a vector space over a field K. Suppose F is a subfield of K. Then V is also a vector space over F! Why? What do you think the scalar multiplication ought to be? (See the notes on page 92 for some remarks on this.)

    As an example of this phenomenon, $\mathbb{R}[x]$, besides being an $\mathbb{R}$–vector space, is also a $\mathbb{Q}$–vector space. Vector addition is the usual addition of polynomials. As for scalar multiplication, when we consider $\mathbb{R}[x]$ as a $\mathbb{Q}$–vector space, we only allow multiplication of polynomials by *rational* numbers—we ignore the fact that we can multiply polynomials by arbitrary real numbers.

    Similarly, $M_2(\mathbb{Q}[\sqrt{2}]$, besides being a $\mathbb{Q}[\sqrt{2}]$–vector space, is also a $\mathbb{Q}$–vector space.

Remark 3.3

Now observe that all these examples of vector spaces have the following properties:

1. For any scalar f, f times the zero vector is just the zero vector.
2. For any vector v, the *scalar* 0 times v is the zero *vector*.
3. For any scalar f and any vector v, $(-f) \cdot v = -(f \cdot v)$.
4. If v is a nonzero vector, then $f \cdot v = 0$ for some scalar f implies $f = 0$.

These properties somehow seem very natural, and one would expect them to hold for *all* vector spaces. Just as in Remark 2.8, where we considered a similar set of properties for rings, we would like these properties to be deducible from the vector space axioms themselves. This would, among other things, convince us that our vector space axioms are the "correct" ones, that is, they yield objects that behave more or less like the examples above instead of objects that are rather pathological. As it turns out, our expectations are not misguided: these properties *are* deducible from the vector space axioms, and therefore *do* hold in all vector spaces. We will leave the verification of this to the exercises (see Exercise 2).

Linear Independence, Bases, Dimension

Now, given a field F and an F-vector space V, it is natural to wonder about the *size* of V. To measure this size, we need to consider the concept of the *dimension* of a vector space.

Let us contrast $\mathbb{R}^2$ with $\mathbb{R}^3$. We all share the intuition that $\mathbb{R}^3$ is somehow *bigger* than $\mathbb{R}^2$. But what precisely is it about $\mathbb{R}^2$ and $\mathbb{R}^3$ that makes us feel that one is bigger than the other? If we examine our intuition a little more closely, we discover that the reason that $\mathbb{R}^3$ seems bigger than $\mathbb{R}^2$ is that $\mathbb{R}^3$ has *three* coordinate axes, while $\mathbb{R}^2$ has only two. *Hidden in this fact is the concept of the dimension of a vector space.* And in fact, without necessarily having paused to think through the notion of dimension or make it precise, most of us have already absorbed this concept and integrated it into our lives—we readily describe $\mathbb{R}^2$ as a *2-dimensional space* and $\mathbb{R}^3$ as a *3-dimensional space.*

With this in mind, what should we take to be the dimension of a vector space? The number of coordinate axes it contains? As it turns out, this is indeed correct, but we have some work to do first. Remember, a vector space is an *algebraic* object. It is defined as an abelian group $(V, +)$ along with a scalar multiplication $F \times V \to V$ with the properties that we have described above. Thus, while the term "coordinate axes" has an obvious meaning in the geometric examples of $\mathbb{R}^2$ and $\mathbb{R}^3$, it is not clear what meaning it should have in a general vector space. So our first task is to convert the geometric notion of coordinate axes into an *algebraic* notion. Next, we need to worry about the possibility that an arbitrary vector space defined purely algebraically may not have any coordinate axes at all, as well as the possibility that different sets of coordinate axes of the same vector space may have different numbers of axes in each set! If either of these possibilities were to occur, we would not have a unique number that we could assign as the dimension of the vector space. As it turns out, neither of these can happen, and our second task is to consider the impossibility of these two scenarios.

Let us turn to the first task. Focusing on $\mathbb{R}^2$ for convenience, let us denote the vector with tip at the point $(1, 0)$ by $\mathbf{i}$, and the one with the tip at the point $(0, 1)$ by $\mathbf{j}$. From vector calculus, we know that if we take an arbitrary vector in $\mathbb{R}^2$, say $\mathbf{u}$, with its tip at (a, b), then the projection of $\mathbf{u}$ onto the x-axis is just a times the vector $\mathbf{i}$ and the projection on the y-axis is just b times the vector $\mathbf{j}$. The parallelogram law then shows that $\mathbf{u}$ is the sum of $a \cdot \mathbf{i}$ and $b \cdot \mathbf{j}$, that is, $\mathbf{u} = a \cdot \mathbf{i} + b \cdot \mathbf{j}$. Since $\mathbf{u}$ was taken to be an arbitrary vector, we find that *every* vector in $\mathbb{R}^2$ can be written as a scalar times $\mathbf{i}$ added to another scalar times $\mathbf{j}$. This example motivates two definitions.

Definition 3.4
A linear combination of vectors $v_1, \cdots, v_n$ is any vector that can be written as $a_1 \cdot v_1 + \cdots + a_n \cdot v_n$ for suitable scalars $a_1, \cdots, a_n$.

Thus, what we found above is that every vector in $\mathbb{R}^2$ can be written as a linear combination of the vectors $\mathbf{i}$ and $\mathbf{j}$. (To give you some specific examples, the vectors $\mathbf{i} + \mathbf{j}$, $\sqrt{2}\mathbf{i} - 3\mathbf{j} = \sqrt{2}\mathbf{i} + (-3)\mathbf{j}$, and $\pi\mathbf{i} + 3\pi^2\mathbf{j}$ are all linear combinations of $\mathbf{i}$ and $\mathbf{j}$.)

The other definition motivated by the example of the vectors $\mathbf{i}$ and $\mathbf{j}$ in $\mathbb{R}^2$ is the following:

Definition 3.5
Given a field F and an F-vector space V, a subset S of V is said to *span* V (or S is said to be a *spanning set for* V) if every vector $v \in V$ can be written as $\sum_{i=1}^{n} a_i \cdot v_i$ for some choice of vectors $v_i \in S$ and some choice of scalars a_i $(i = 1, \ldots, n)$. In other words, in the language of

Definition 3.4 above, S is a spanning set for V if every vector in V is expressible as a *linear combination* of elements of S.

The discussion before Definition 3.4 showed that the set $S = \{\mathbf{i}, \mathbf{j}\}$ is a spanning set for $\mathbb{R}^2$ (why?). Here is another example:

Example 3.6
We have seen in Example 3.2.5 that $\mathbb{Q}[\sqrt{2}]$ is a $\mathbb{Q}$–vector space. Note that every element of $\mathbb{Q}[\sqrt{2}]$ is of the form $a + b\sqrt{2}$ for suitable a and $b \in \mathbb{Q}$. Thinking of "a" as "$a \cdot 1$," this tells us that every element of $\mathbb{Q}[\sqrt{2}]$ is expressible as a linear combination of 1 and $\sqrt{2}$. (We are thinking of 1 as a vector in this last statement. Recall the discussion of the dual role of $\mathbb{Q}$ in Example 3.2.5.) Hence, $S = \{1, \sqrt{2}\}$ is a spanning set for the $\mathbb{Q}$ vector space $\mathbb{Q}[\sqrt{2}]$.

So, returning to our study of dimension, should we take the algebraic analog of coordinate axes to be any set of vectors $v_1, \ldots, v_n$ that span V? No, not yet! We need to consider one more fact. Going back to $\mathbb{R}^2$, let us write $\mathbf{w}$ for the vector with tip at $(1/\sqrt{2}, 1/\sqrt{2})$. Then $\mathbf{i}$, $\mathbf{j}$, and $\mathbf{w}$ also span $\mathbb{R}^2$. (This is very trivial to see—the vector with tip at (a, b) can be written as the sum $a \cdot \mathbf{i} + b \cdot \mathbf{j} + 0 \cdot \mathbf{w}$. Can you show that it can also be written as $(a - 1/\sqrt{2}) \cdot \mathbf{i} + (b - 1/\sqrt{2}) \cdot \mathbf{j} + \mathbf{w}$?) Yet, we do not think of $\mathbf{i}$, $\mathbf{j}$, and $\mathbf{w}$ *together* as a set of coordinate axes—there is redundancy in this set. Any two of $\mathbf{i}$, $\mathbf{j}$, and $\mathbf{w}$ already span $\mathbb{R}^2$. (If you are not familiar with the fact that $\mathbf{i}$ and $\mathbf{w}$ also span $\mathbb{R}^2$, we will see this in a moment, but you should also do Exercise 3.) How should we choose a spanning set that avoids this sort of redundancy? To gain some insight, let us examine why any two of $\mathbf{i}$, $\mathbf{j}$, and $\mathbf{w}$, say $\mathbf{i}$ and $\mathbf{w}$, would do as a spanning set for $\mathbb{R}^2$. Observe that $\mathbf{j} = -\mathbf{i} + \sqrt{2}\mathbf{w}$. Thus, any linear combination $a \cdot \mathbf{i} + b \cdot \mathbf{j} + c \cdot \mathbf{w}$ of $\mathbf{i}$, $\mathbf{j}$, and $\mathbf{w}$ can be written as $(a - b) \cdot \mathbf{i} + (c + \sqrt{2}b) \cdot \mathbf{w}$ by simply substituting $-\mathbf{i} + \sqrt{2}\mathbf{w}$ for $\mathbf{j}$. In other words, every linear combination of $\mathbf{i}$, $\mathbf{j}$, and $\mathbf{w}$ can be *rewritten* as a linear combination of $\mathbf{i}$ and $\mathbf{w}$. This gives us a clue to the redundancy, and in turn, yields a very important concept.

In the context of a general vector space V, let us call a vector v_i in a spanning set $\{v_1, \ldots, v_n\}$ *redundant* if the subset obtained by removing v_i, that is, the set $\{v_1, \ldots, v_{i-1}, v_{i+1}, \ldots, v_n\}$, is itself a spanning set for V. (In other words, every vector in V can already be expressed as a linear combination of the vectors $v_1, \ldots, v_{i-1}, v_{i+1}, \ldots, v_n$, and the vector v_i is not needed.) We will say that there is redundancy in the spanning set $\{v_1, \ldots, v_n\}$ if any one of the vectors in this set is redundant.

Now, what the example of the vectors $\mathbf{i}$, $\mathbf{j}$, and $\mathbf{w}$ in $\mathbb{R}^2$ shows is the following: Suppose a spanning set $\{v_1, \ldots, v_n\}$ in a general vector space V is such that one of the vectors, say v_i, can be written as a linear combination of the remaining vectors $v_1, \ldots, v_{i-1}, v_{i+1}, \ldots, v_n$. *The vector v_i is then redundant.* All we have to do to see this is to generalize the calculations in the example above: Suppose that $v_i = a_1 \cdot v_1 + \cdots + a_{i-1} \cdot v_{i-1} + a_{i+1} \cdot v_{i+1} + \cdots a_n \cdot v_n$ for some scalars $a_1, \ldots, a_{i-1}, a_{i+1}, \ldots, a_n$. Then, given a linear combination $c_1 \cdot v_1 + \cdots + c_n \cdot v_n$ of the vectors $v_1, \ldots, v_n$, simply plug in $a_1 \cdot v_1 + \cdots + a_{i-1} \cdot v_{i-1} + a_{i+1} \cdot v_{i+1} + \cdots a_n \cdot v_n$ for v_i in this expression to obtain the linear combination $(c_1 + c_i a_1)v_1 + (c_2 + c_i a_2)v_2 + \cdots + (c_{i-1} + c_i a_{i-1})v_{i-1} + (c_{i+1} + c_i a_{i+1})v_{i+1} + \cdots + (c_n + c_i a_n)v_n$. Therefore, given any vector in V, we can first write it as a linear combination of $v_1, \ldots, v_n$ (which can be done since the set $\{v_1, \ldots, v_n\}$ is a spanning set for V), and then using the method above, we can rewrite it as a linear combination of $v_1, \ldots, v_{i-1}, v_{i+1}, \ldots, v_n$. Thus, the set $\{v_1, \ldots, v_{i-1}, v_{i+1}, \ldots, v_n\}$ already spans V—we do not need the vector v_i at all!

On the other hand, if v_i is redundant, this of course means that the set $\{v_1, \ldots, v_{i-1}, v_{i+1}, \ldots, v_n\}$ spans V. Every vector in V is therefore expressible (by definition) as a linear combination of the vectors $v_1, \ldots, v_{i-1}, v_{i+1}, \ldots, v_n$. In particular, v_i is also expressible as a linear combination of $v_1, \ldots, v_{i-1}, v_{i+1}, \ldots, v_n$. Putting this together with the discussion in the previous paragraph, we find that any one vector in the spanning set $\{v_1, v_2, \ldots, v_n\}$ is redundant *if and only if* it is expressible as a linear combination of the remaining vectors.

Now let us try to express this condition in a slightly different way. Suppose one of the vectors, say v_i, is expressible as a linear combination of the remaining vectors. Thus, $v_i = a_1 \cdot v_1 + \cdots + a_{i-1} \cdot v_{i-1} + a_{i+1} \cdot v_{i+1} + \cdots a_n \cdot v_n$ for some scalars $a_1, \ldots, a_{i-1}, a_{i+1}, \ldots, a_n$. Bringing v_i to the other side, we have $a_1 \cdot v_1 + \cdots + a_{i-1} \cdot v_{i-1} + (-1) \cdot v_i + a_{i+1} \cdot v_{i+1} + \cdots a_n \cdot v_n = 0$. (Observe that the scalar multiplying v_i in the expression above is not equal to zero.) Conversely, suppose there exist scalars $a_1, \ldots, a_n$, *not all zero,* such that $a_1 \cdot v_1 + \cdots + a_n \cdot v_n = 0$. At least one of these scalars is nonzero; let us assume it is a_i. We may thus divide by a_i, so we can write $v_i = (-a_1/a_i)\cdot v_1 + \cdots + (-a_{i-1}/a_i)\cdot v_{i-1} + (-a_{i+1}/a_i)\cdot v_{i+1} + \cdots (-a_n/a_i)\cdot v_n$. What this paragraph shows is that some vector in the set $\{v_1, \ldots, v_n\}$ can be expressed as a linear combination of the remaining vectors *if and only if* there exist scalars $a_1, \ldots, a_n$, not all zero, such that $a_1 \cdot v_1 + \cdots + a_n \cdot v_n = 0$. Combining this with the discussions of the previous paragraph, we find that we have proved the following:

Lemma 3.7
Let V be a vector space over a field F, and let $S = \{v_1, \ldots, v_n\}$ be a spanning set for V. Then, the following are equivalent:

1. *There is redundancy in S,*
2. *Some vector in S is expressible as a linear combination of the remaining vectors in S, and*
3. *There exist scalars $a_1, \ldots, a_n$, not all zero, such that $a_1 \cdot v_1 + \cdots + a_n \cdot v_n = 0$.*

With this lemma in mind, we make the following definition:

Definition 3.8
Let F be a field and V an F-vector space. Let $v_1, \ldots, v_n$ be elements of v. Then $v_1, \ldots, v_n$ are said to be *linearly dependent over F*, or *F-linearly dependent* if there exist scalars $a_1, \ldots, a_n$, not all zero, such that $a_1 \cdot v_1 + \cdots + a_n \cdot v_n = 0$. If no such scalars exist, then $v_1, \ldots, v_n$ are said to be *linearly independent over F*, or *F-linearly independent*. (If there is no ambiguity about the field F, the vectors are merely referred to as *linearly dependent* or *linearly independent*. Also, if $v_1, \ldots, v_n$ are linearly independent, respectively linearly dependent, vectors, then the *set* $\{v_1, \ldots, v_n\}$ is said to be a linearly independent, respectively a linearly dependent, set.) An arbitrary subset S of V is said to be *linearly independent* if every finite subset of S is linearly independent. Similarly, an arbitrary subset S of V is said to be *linearly dependent* if some finite subset of S is linearly dependent.

(Can you show that if v is a *nonzero* vector, then the set $\{v\}$ must be linearly independent? See Remark 3.3.4.)

Thus, the implications $1 \Leftrightarrow 3$ of Lemma 3.7 can be stated in this new language as follows: There is redundancy in S *if and only if* S is linearly dependent. Also, let us illustrate the meaning of the last two sentences of the definition above. Let us consider the following:

Example 3.9

We saw in Example 3.2.7 that $\mathbb{R}[x]$ is a vector space over $\mathbb{R}$. Now consider the set $S = \{1, x, x^2, x^3, \ldots\}$. This is, of course, an *infinite* set. Consider any nonempty finite subset of S, for instance, the subset $\{x, x^5, x^{17}\}$ or the subset $\{1, x^3, x^{99}, x^{100}, x^{1001}, x^{1004}\}$ or the subset $\{1, x, x^2, x^{20}\}$. In general, a nonempty finite subset of S would contain n elements (for some $n \geq 1$), and these elements would be various powers of x—say x^{i_1}, x^{i_2}, ..., x^{i_n}. These elements are definitely linearly independent, since if $a_1 x^{i_1} + \cdots + a_n x^{i_n}$ is the zero polynomial, then by the definition of the zero polynomial, each a_i must be zero. *This is true regardless of which finite subset of S we take*—all that would be different in different finite subsets is the number of elements (the integer n) and the particular powers of x (the integers i_1 through i_n) chosen. Thus, according to our definition, the set S is linearly independent.

On the other hand, consider the subset $T = S \cup \{1 + x\}$. Any finite subset of T that does not contain all three vectors 1, x and $1 + x$ will be linearly independent (check!). However, this alone is not enough for you to conclude that T is a linearly independent set. For the subset $\{1, x, 1 + x\}$ of T is linearly dependent: $1 \cdot 1 + 1 \cdot x + (-1) \cdot (1 + x) = 0$. By the definition above, T is a linearly *dependent* set.

We are now ready to construct the algebraic analog of coordinate axes. We will choose as our candidate any set of vectors that spans our vector space and in which there is no redundancy. Moreover, instead of using the term coordinate axes (which is inspired by the geometric examples of $\mathbb{R}^2$ and $\mathbb{R}^3$), we will coin a new term—the algebraic analog of coordinate axes will now be called a *basis*. Since redundancy is equivalent to linear dependence (Lemma 3.7), lack of redundancy is equivalent to linear *in*dependence. We hence have the following definition:

Definition 3.10

Let F be a field and V an F-vector space. A subset S of V is said to be a *basis* of V if S spans V and there is no redundancy in S. (As remarked above, lack of redundancy is equivalent to linear independence, so we can alternatively define S to be a basis of V if S spans V and is linearly independent.) The individual vectors that belong to S are referred to as *basis vectors*. Sometimes, when we wish to emphasize the field of scalars, we refer to S as an F-*basis* of V.

Here are some examples of bases of vector spaces:

Examples 3.11

1. The set consisting of the vectors $\mathbf{i}$ and $\mathbf{j}$ is a basis for $\mathbb{R}^2$. We have already seen in the text that $\mathbf{i}$ and $\mathbf{j}$ span $\mathbb{R}^2$. Now, if $a\mathbf{i} + b\mathbf{j} = 0$ for some scalars a and b, then this means that the vector with the tip at (a, b) is the zero vector. Hence, (a, b) must equal $(0, 0)$, so equating x and y components, a and b must both be zero. Thus, there cannot exist real numbers a and b, *not both zero,* such that $a\mathbf{i} + b\mathbf{j} = 0$, so by definition, $\mathbf{i}$ and $\mathbf{j}$ are linearly independent.

   Can you show that the set consisting of the vectors $\mathbf{i}$ and $\mathbf{w} = (1/\sqrt{2}, 1/\sqrt{2})$ also forms a basis? (We have already seen in the text that $\mathbf{i}$ and $\mathbf{w}$ span $\mathbb{R}^2$. Thus, you only need to show that $\mathbf{i}$ and $\mathbf{w}$ are linearly independent. For this, consider an arbitrary linear combination $a\mathbf{i} + b\mathbf{w}$, where a and b are scalars, and note that if $a\mathbf{i} + b\mathbf{w}$ equals the zero vector, then the x and y components of $a\mathbf{i} + b\mathbf{w}$ must each equal zero. The x component of $a\mathbf{i} + b\mathbf{w}$ is $a + b/\sqrt{2}$, and the y component is $b/\sqrt{2}$. So?)
2. The set consisting of the vectors with tips at $(1, 0, 0)$, $(0, 1, 0)$, and $(0, 0, 1)$ forms a basis for $\mathbb{R}^3$. (Why?) Can you show that the set consisting of the vectors with tips at $(1, 0, 0)$, $(1/\sqrt{2}, 1/\sqrt{2}, 0)$, and $(0, 0, 1)$ also forms a basis? (You want to show two things: first, that given any vector (a, b, c) in $\mathbb{R}^3$, there exist scalars s, t, and u such that $(a, b, c) = s(1, 0, 0) + t(1/\sqrt{2}, 1/\sqrt{2}, 0) + u(0, 0, 1)$, and second, that a linear combination such as $s(1, 0, 0) + t(1/\sqrt{2}, 1/\sqrt{2}, 0) + u(0, 0, 1)$ gives the zero vector (that is, the vector $(0, 0, 0)$) only if s, t, and u are all zero. For the first, compare the x, y, and z coordinates of the vectors on both sides to derive three linear equations in the three variables s, t, and u, and show that this system of equations has a solution. As for the second, again compare x, y, and z coordinates, this time on both sides of the equation $(0, 0, 0) = s(1, 0, 0) + t(1/\sqrt{2}, 1/\sqrt{2}, 0) + u(0, 0, 1)$, and derive three linear equations in the variables s, t, and u. Show that this second system of equations has precisely one solution: $s = t = u = 0$.)

3. The set consisting of the elements 1 and $\sqrt{2}$ forms a basis for $\mathbb{Q}[\sqrt{2}]$ as a vector space over $\mathbb{Q}$. (We have seen in Example 3.6 above that 1 and $\sqrt{2}$ span $\mathbb{Q}[\sqrt{2}]$. As for the $\mathbb{Q}$-linear independence of 1 and $\sqrt{2}$, you were asked to prove this in Exercise 4 in Chapter 2!)
4. The set $\{1, x, x^2, \ldots\}$ forms a basis for $\mathbb{R}[x]$ as a vector space over $\mathbb{R}$. It is clear that this set spans $\mathbb{R}[x]$, since every polynomial is a sum of terms of the form $a_i x^i$ for suitable $a_i \in \mathbb{R}$ and suitable i; in other words, every polynomial is an "$\mathbb{R}$-linear combination" of the "vectors" x^i for various i. As for the linear independence, see the argument in Example 3.9 above. What can you say about the set $\{1, 1 + x, (1 + x)^2, \ldots\}$? (See Exercise 4.)
5. Consider $F_n[x]$ as an F-vector space (see Example 3.2.8 above). You should easily be able to describe a basis for this space.
6. The set $\{1, \sqrt{2}, \sqrt{3}, \sqrt{6}\}$ forms a basis for $\mathbb{Q}[\sqrt{2}, \sqrt{3}]$ as a vector space over $\mathbb{Q}$. (Why does this set span $\mathbb{Q}[\sqrt{2}, \sqrt{3}]$? Note that you were asked to prove the $\mathbb{Q}$-linear independence of this set in Exercise 7 in Chapter 2!)
7. Can you prove that the n^2 matrices $e_{i,j}$ (see Exercise 6 of Chapter 2 for this notation) are a basis for $M_n(\mathbb{R})$? To start you off, here is a hint: In $M_2(\mathbb{R})$, for example, a matrix such as

$$\begin{pmatrix} 1 & 2 \\ 3 & 4 \end{pmatrix}$$

can be written as the linear combination

$$e_{1,1} + 2e_{1,2} + 3e_{2,1} + 4e_{2,2}.$$

8. Going back to $\mathbb{Q}[\sqrt{2}]$, show that the vectors 1 and $1 + \sqrt{2}$ also form a basis. (Hint: Any vector $a + b\sqrt{2}$ can be rewritten as $(a - b) + b(1 + \sqrt{2})$. So?) Now show that if V is any vector space over any field with basis $\{v_1, v_2\}$, then the vectors $v_1, v_1 + v_2$ also form a basis. How would you generalize this to a vector space that has a basis consisting of three elements $\{v_1, v_2, v_3\}$?
9. Consider the vector space F^∞ of Example 3.2.3 above. You may find it hard to describe explicitly a basis for this space. However, let e_i (for $i = 0, 1, \ldots$) be the infinite-tuple with 1 in the ith position and zeros elsewhere. (Thus, $e_0 = (1, 0, 0, \ldots)$, $e_1 = (0, 1, 0, \ldots)$, etc.) Why is the set $S = \{e_0, e_1, e_2, \ldots\}$ *not* a ba-

sis for F^{∞}? Is S at least linearly independent? (See the notes on page 92 for some comments on this example.)

Now that we have arrived at the algebraic analog of coordinate axes, we turn our attention to the next step in our program—we need to show that every vector space has a basis, and that different bases of the same vector space have the same number of elements in them.

The first of these two tasks, namely, showing that every vector space has a basis, is a little tricky to do with just the background in set theory and logic that we have at this point. To do full justice to this task, we need to invoke *Zorn's Lemma*, which is an extremely useful tool of logic. (Zorn's Lemma, in spite of its name, is really not a lemma, but an *axiom* of logic. See the remarks on this on page 93 in the notes.) Without Zorn's Lemma at our disposal, how should we convince ourselves that every vector space has a basis?

We will do two things. First, we will actually present a full proof that every vector space has a basis. The problem with this proof is that at a criticial juncture, our presentation will simply invoke Zorn's Lemma, and if you do not know Zorn's Lemma, this of course is not very illuminating. The advantage of presenting this proof, however, is that you will at least see the rest of the arguments for the existence of a basis, and as well, this proof may motivate you to learn about Zorn's Lemma yourselves. The other thing that we will do is to give a proof of the existence of a basis in a special case, namely, when we *know* that the vector space in question has a finite spanning set. Since many of the vector spaces that we consider in this book (although definitely not all) fall into this category, we can be assured that at least in these vector spaces, we will be able to find bases.

Let us consider the special case first.

Theorem 3.12
Let V be a vector space over a field F. Let S be a spanning set for V, and assume that S is a finite set. Then some subset of S is a basis of V. In particular, every vector space with a finite spanning set has a basis.

Proof If the zero vector appears in S, then the set $S' = S - \{0\}$ that we get by throwing out the zero vector will still span V and will still be finite (why?). Any subset of S' will also be a subset of S, so if we can show that some subset of S' must be a basis of V, then we would have proved that some subset of our original set S must be a basis of V. Hence, we may assume that we are given a spanning set S for V that is not only finite, but one in which none of the vectors is zero.

Let $S = \{v_1, v_2, \ldots, v_n\}$ for some $n \geq 1$. If there is no redundancy in S, then there is nothing to prove: S would be a basis by the very definition of basis. So assume that there is redundancy in S. By relabeling if necessary, we may assume that v_n is redundant. Thus, $S_1 = \{v_1, v_2, \ldots, v_{n-1}\}$ is itself a spanning set for V. Once again, if there is no redundancy in S_1, then we would be done; S_1 would be a basis. So assume that there is redundancy in S_1. After relabeling if necessary, we may assume that v_{n-1} is redundant. Thus, $S_2 = \{v_1, v_2, \ldots, v_{n-2}\}$ is itself a spanning set for V. Again, if there is no redundancy in S_2, we would be done; S_2 would be a basis. If not ... This process must stop somewhere, since at worst, we would shrink our spanning set down to one vector, say $S_{n-1} = \{v_1\}$, and a set containing just one nonzero vector must be linearly independent (why?); so S_{n-1} would form a basis. (Note that this is only the worst case; in actuality, this process may stop well before we shrink our spanning set down to just one vector.) When this process stops, we would have a subset of S that would be a basis of V. □

Remark 3.13

Notice that to prove that bases exist (in the special case where V has a finite spanning set) what we really did was to show that every finite spanning set of V can be shrunk down to a basis of V. This result is true more generally: Given *any* spanning set S of a vector space V (in other words, not just a finite spanning set S), there exists a subset S' of S that forms a basis of V. We will not prove this here.

Now we will present the proof that bases exist in all vector spaces, not just in those with a finite spanning set. Recall from the discussions before Theorem 3.12 our caveat about this proof: at a crucial step, it invokes Zorn's Lemma.

Theorem 3.14

Every vector space has a basis.

Proof (Sketch:) Given a vector space V and a linearly independent subset S of V, we call S a *maximal linearly independent subset* if the set $S \cup \{v\}$ is linearly dependent for every vector v such that $v \notin S$. In $\mathbb{R}^2$, for instance, the set $S = \{\mathbf{i}, \mathbf{j}\}$ is a maximal linearly independent subset, since if $\mathbf{u}$ is *any* vector in $\mathbb{R}^2$ with $\mathbf{u} \neq (1, 0)$ and $\mathbf{u} \neq (0, 1)$, then $S \cup \{\mathbf{u}\}$ is linearly dependent (why?).

Notice that if you know that a given vector space V has a basis B, then B must necessarily be a maximal linearly independent subset of V. (Why?)

By Zorn's Lemma applied to the set of all subsets of V consisting of linearly independent elements of V (see the remarks on page 93 in the notes), V has a maximal linearly independent set. We will show that any maximal linearly independent subset of V will be a basis. Accordingly, let T be a maximal linearly independent subset. Since T is already linearly independent, we only need to show that T spans V. So let v be any nonzero vector in V: we need to show that v can be written as a linear combination of elements of T. If v is already in T, there is nothing to prove (why?). If v is not in T, then since T is a maximal linearly independent subset, $T \cup \{v\}$ is linearly dependent. Thus, there exists a relation $f_0 v + f_1 t_1 + f_2 t_2 + \cdots f_k t_k = 0$ for some scalars f_0, f_1, $\cdots$, f_k (not all zero), and some vectors t_1, t_2, $\cdots$, t_k of T. Notice that $f_0 \neq 0$, since otherwise, our relation would read $f_1 t_1 + f_2 t_2 + \cdots f_k t_k = 0$ (with not all f_i equal to zero), which is impossible since the t_i are in T and T is a linearly independent set. Therefore, we can divide by f_0 to find $v = (-f_1/f_0)t_1 + (-f_2/f_0)t_2 + \cdots + (-f_k/f_0)t_k$. Hence v can be written as a linear combination of elements of T, so T spans V.

Thus, any maximal linearly independent subset of V (which exists by Zorn's Lemma) is a basis of V. □

Having proved (after a fashion!) that every vector space has a basis, we now need to show that different bases of a vector space have the same number of elements in them. (Remember our original program. We wish to measure the size of a vector space, and based on our examples of $\mathbb{R}^2$ and $\mathbb{R}^3$, we think that a good measure of the size would be the number of coordinate axes, or basis elements, that a vector space has. However, for this to make sense, we need to be guaranteed that every vector space has a basis—we just convinced ourselves of this—and that different bases of a vector space have the same number of elements in them.) In preparation, we will prove an important lemma. Our desired results will fall out as corollaries.

Lemma 3.15
Let V be a vector space over a field F, and let $B = \{v_1, \ldots, v_n\}$ $(n \geq 1)$ be a basis for V. Let $C = \{w_1, \ldots, w_m\}$ be a linearly independent set. Then $m \leq n$.

Proof The basic idea behind the proof is to replace vectors in the basis B one after another with vectors in C, and observing at the end that if m were greater than n, then there would not be enough replacements of elements of B to guarantee linear independence of the set C.

We begin as follows: Since B spans V, every vector in V is expressible as a linear combination of elements of B. In particular, we may write w_1 as a linear combination of elements of B, that is, $w_1 = c_1v_1 + +c_2v_2 + \ldots + c_nv_n$ for suitable scalars c_i, not all zero. Since one of these scalars is nonzero, we may assume for convenience (by relabeling the vectors of B if necessary), that $c_1 \neq 0$. As usual, we may write $v_1 = (-1/c_1)w_1 + (-c_2/c_1)v_2 + (-c_3/c_1)v_3 + \ldots + (-c_n/c_1)v_n$. Now go back and study how we proved $1 \Leftrightarrow 2$ in Lemma 3.7. We are going to use the same sort of an argument here: we will prove that the set $\{w_1, v_2, v_3, \ldots, v_n\}$ spans V. For given any vector in V, it can be written as a linear combination $f_1v_1 + f_2v_2 + \ldots + f_nv_n$ for suitable scalars f_i (why?). Now, in this expression, substitute $(-1/c_1)w_1 + (-c_2/c_1)v_2 + (-c_3/c_1)v_3 + \ldots + (-c_n/c_1)v_n$ for v_1, and what do you get?—a linear combination of w_1, v_2, v_3, …, v_n! Thus, the set $\{w_1, v_2, v_3, \ldots, v_n\}$ spans V as claimed.

Now observe what we have done: we have replaced v_1 with w_1. Let us take this to the next step. Since the set $\{w_1, v_2, v_3, \ldots, v_n\}$ spans V, we can write w_2 as a linear combination of elements of this set. Thus, $w_2 = g_1w_1 + g_2v_2 + g_3v_3 + \ldots + g_nv_n$ for suitable scalars g_i, not all zero. Now the scalars g_2, g_3, …, g_n cannot all be zero, since g_1 would then have to be nonzero (why?) and this relation would then read $w_2 = g_1w_1$—a contradiction, as the set C is linearly independent. Hence, one of the scalars $g_2, g_3, \ldots, g_n$ must be nonzero. For convenience, we may assume (by relabeling the vectors v_2, v_3, …, v_n if necessary) that $g_2 \neq 0$. Dividing by g_2 and moving all terms but v_2 to one side, we can write v_2 as a linear combination of the vectors $w_1, w_2, v_3, \ldots, v_n$. Exactly as in the last paragraph, we find that since the set $\{w_1, v_2, v_3, \ldots, v_n\}$ spans V, the set $\{w_1, w_2, v_3, \ldots, v_n\}$ also spans V.

So far, we have succeeded in replacing v_1 with w_1 and v_2 with w_2, and the resultant set $\{w_1, w_2, v_3, \ldots, v_n\}$ still spans V. Now continue this process, and consider what would happen if we were to assume that m is greater than n. Well, we would replace v_3 by w_3, v_4 by w_4, etc., and then v_n by w_n. (We know that we would be able to replace all the v's with w's because by assumption, there are more w's than v's.) At each stage of the replacement, we would be left with a set that spans V. In particular, the set we would be left with after replacing v_n by w_n, namely $\{w_1, w_2, \ldots, w_n\}$, would span V. But since we assumed that m is greater than n, there would be at least one "w" left, namely w_{n+1}. Since $\{w_1, w_2, \ldots, w_n\}$ would span V, we would be able to write w_{n+1} as a linear combination of the vectors w_1, w_2, …, w_n. This is a contradiction, since the set C is linearly independent! Hence m cannot be greater than n, that is, $m \leq n$! □

We are now ready to prove that different bases of a given vector space have the same number of elements. We will distinguish between two cases: vector spaces having bases with finitely many elements, and those having bases with infinitely many elements. We will take care of the infinite case first.

Corollary 3.16
If a vector space V has one basis with an infinite number of elements in it, then every other basis of the vector space also has an infinite number of elements in it.

Proof Let S be the basis of V with an infinite number of elements (that exists by hypothesis), and let T be any other basis. Assume that T has only finitely many elements in it, say m. Since S has infinitely many elements in it, we can certainly pick $m + 1$ vectors from it. So pick any $m + 1$ vectors from S and denote this selected set of vectors by S'. Since the vectors in S' are part of the basis S, they are certainly linearly independent. We may think of the set T as the set "B" of Lemma 3.15 (after all, T is a basis), and we may think of the set S' as the set "C" of the same lemma (after all, S' is linearly independent).

The lemma then shows that $m + 1 \leq m$, a clear contradiction. Hence T must also be infinite! □

We settle the finite case now.

Corollary 3.17
If a vector space V has one basis with a finite number of elements n, then every basis of V contains n elements.

Proof Let $S = \{x_1, \ldots, x_n\}$ be the basis of V with n elements in it (that exists by hypothesis), and let T be any other basis. If T were infinite, Lemma 3.16 above says that S must also be infinite. Since this is not true, we find that T must have a finite number of elements in it. So, assume that T has m elements in it, say $T = \{y_1, \ldots, y_m\}$. We wish to show that $m = n$. We may think of S as the set "B" of Lemma 3.15. Also, we may think of the set T as the set "C" of the lemma, since T, being a basis, is certainly linearly independent. Then the lemma says that m must be less than or equal to n. Now let us reverse this situation: let us think of T as the set "B," and let us think of S as the set "C". (Why can we do this?) Then the lemma says that n must be less than or equal to m. Thus, we have $m \leq n$ and $n \leq m$, so we find that $n = m$. □

We are finally ready to make the notion of the size of a vector space precise!

Definition 3.18
A vector space V over a field F is said to be *finite-dimensional* (or *finite-dimensional over F*) if it has a basis with a finite number of elements in it; otherwise, it is said to be *infinite-dimensional* (or *infinite-dimensional over F*). If V is finite-dimensional, the *dimension* of V is defined to be the number of elements in any basis. If V is infinite-dimensional, the *dimension* of V is defined to be infinite. If V has dimension n, then V is also referred to as an *n-dimensional* space (or as being *n-dimensional over F*).

Let us consider the dimensions of some of the vector spaces in Examples 3.2 (see also Examples 3.11, where we consider bases of these vector spaces). $\mathbb{R}^2$ and $\mathbb{R}^3$ have dimensions 2 and 3 (respectively) as vector spaces over $\mathbb{R}$. What is the dimension of $\mathbb{R}^n$? $\mathbb{Q}[\sqrt{2}]$ is 2-dimensional over $\mathbb{Q}$. $\mathbb{R}[x]$ is infinite-dimensional over $\mathbb{R}$, while $\mathbb{Q}[\sqrt{2}, \sqrt{3}]$ is 4-dimensional over $\mathbb{Q}$. Similarly, $M_n(\mathbb{R})$ is n^2-dimensional over $\mathbb{R}$. What is the dimension of $F_n[x]$ over F? (Warning! It is *not* n.)

The following result is crucial:

Theorem 3.19

Let V be an n-dimensional vector space. Then every subset S of V consisting of more than n elements is linearly dependent.

Proof Assume, to the contrary, that V contains a linearly independent subset S that contains more than n elements. Therefore, we can find $n + 1$ distinct elements in S. Call them $v_1, v_2, \ldots, v_{n+1}$ and write C for the set $\{v_1, v_2, \ldots, v_{n+1}\}$. Let B be any basis. By the very definition of dimension, B must have n elements. Now apply Theorem 3.6 to the sets B and C—we find that $n + 1 \leq n$, which is a contradiction. Hence every subset of V consisting of more than n elements must be linearly dependent. □

We will state one more theorem that is of fundamental importance. Once again, this is a theorem that holds even when V is not assumed to be finite-dimensional, but a full proof requires the use of Zorn's Lemma. We will omit the proof of the general case (see the remarks on page 94 in the notes), and only prove the theorem under the assumption that V is finite-dimensional.

Theorem 3.20

Let V be a finite-dimensional vector space, and let C be a linearly indepenent set. Then C can be expanded to a basis of V, that is, there exists a basis B of V such that $C \subseteq B$.

Proof Let n be the dimension of V. Note that the number of elements in C is at most n, for by Theorem 3.19 above, if C had more than n elements, C would be linearly dependent. So suppose that C has t elements, for some integer $t \leq n$, and suppose that the elements of C are $\{v_1, v_2, \ldots, v_t\}$. If C already spans V, then C would be a basis and we would be done. (And if this happens, you know that t must equal n by Corollary 3.17!) So assume that C does not span V. By the very definition of what it means to span a vector space, there must be a vector in V, call it v_{t+1}, that is not expressible as a linear combination of the elements in C. We claim that the set $C_1 = \{v_1, v_2, \ldots, v_t, v_{t+1}\}$ must be linearly independent. For suppose

$f_1v_1 + \ldots + f_tv_t + f_{t+1}v_{t+1} = 0$ for some scalars f_i, not all of which are zero. Then f_{t+1} cannot be zero, since otherwise our relation would read $f_1v_1 + \ldots + f_tv_t = 0$ for nonzero scalars f_i, and this would violate the linear independence of C. Therefore, we may divide our original relation by f_{t+1} to find $v_{t+1} = (-f_1/f_{t+1})v_1 + \ldots + (-f_t/f_{t+1})v_t$, contradicting the fact that v_{t+1} is not expressible as a linear combination of elements of C. Thus, C_1 is indeed linearly independent as claimed.

Note that the set C_1 has $t + 1$ elements. If C_1 spans V, then C_1 would be a basis of V containing C, and we would be done. So suppose C_1 does not span V; then there must be a vector in V, call it v_{t+2}, that cannot be expressed as a linear combination of elements of C_1. Exactly as in the case with C_1 in the paragraph above, we can show that the set $C_2 = \{v_1, v_2, \ldots, v_t, v_{t+1}, v_{t+2}\}$ must be linearly independent. If C_2 spans V, then C_2 would be a basis of V containing C, and we would be done. If not, there must be a vector in V, call it v_{t+3}, that cannot be expressed as a linear combination of elements of C_2, so the set $C_3 = \{v_1, v_2, \ldots, v_t, v_{t+1}, v_{t+2}, v_{t+3}\}$ must be linearly independent. If C_3 spans V, then C_3 would be a basis of V containing C; if not, we repeat the process ...

Notice that in the process above, we start with our set C with t elements, and at each stage, we come up with a set that has one more element than the set at the previous stage. When we reach a set with exactly n elements, this set *must* span V, for if not, the set we would get at the next stage would contain $n + 1$ elements and would be linearly independent, contradicting Theorem 3.19 above. This set with exactly n elements would therefore be a basis of V containing C. □

Subspaces

The idea behind subspaces is very similar to the idea behind subrings.

Definition 3.21
Given a vector space V over a field F, a *subspace* of V is a subset W of V that is closed with respect to vector addition and scalar multiplication, such that with respect to this addition and scalar multiplication,

W is itself a vector space (that is, W satisfies all the axioms of a vector space).

Now, we saw in the context of rings (Exercise 3 in Chapter 2) that one could have a subset S of a ring R such that S is closed with respect to addition and multiplication, and yet S is *not* a subring of R. It turns out that in the case of vector spaces, it is *enough* for a (nonempty) subset W of a vector space V to be closed with respect to vector addition and scalar multiplication—W will then automatically satisfy all the axioms of a vector space. This is the content of the following theorem.

Theorem 3.22
Let V be a vector space over a field F, and let W be a nonempty subset of V that is closed with respect to vector addition and scalar multiplication. Then W is a subspace of V.

Proof We need to check that all the axioms of a vector space hold. Let us first check that $(W, +)$ is an abelian group. Vector addition in W is associative, since for any $v_1, v_2, v_3 \in W$, we may consider v_1, v_2 and v_3 to be elements of V, and in V, the relation $(v_1 + v_2) + v_3 = v_1 + (v_2 + v_3)$ certainly holds. Next, given any $v \in W$, let us show that $-v$ is also in W. For this we invoke that fact that W is closed with respect to scalar multiplication—since $v \in W$, $-1 \cdot v$ is also in W, and $-1 \cdot v$ is, of course, just $-v$ (see Remark 3.3 above). Now let us show that 0 is in W. Observe that so far, we have not used the hypothesis that W is nonempty. (The proofs that we have given for the fact that addition in W is associative and that every element in W has its additive inverse in W hold *vacuously* even in the case where W is empty. For instance, the chain of arguments $v \in W \Rightarrow -1 \cdot v \in W$ (as W is closed with respect to scalar multiplication) $\Rightarrow -v \in W$ is correct even when there is no vector v in W to begin with!) Now let us use the fact that W is nonempty. Since W is nonempty, it contains at least one vector, call it v. Then, by what we proved above, $-v$ is also in W. Since W is closed under vector addition, $v + (-v)$ is in W, and so 0 is in W. We have thus shown that $(W, +)$ is an abelian group.

It remains to be shown that the four axioms of scalar multiplication also hold for W. But for any r and s in F and v and w in

W, we may consider v and w to be elements of V, and as elements of V, we certainly have the relations $r \cdot (v + w) = r \cdot v + r \cdot w$, $(r + s) \cdot v = r \cdot v + s \cdot v$, $(rs) \cdot v = r \cdot (s \cdot v)$, and $1 \cdot v = v$. Hence, the axioms of scalar multiplication hold for W.

This proves that W is a subspace of V. □

Here are some examples of subspaces. In each case, check that the conditions of Theorem 3.22 apply.

Examples 3.23

1. If you think of $\mathbb{R}^2$ as the vectors lying along the xy plane of 3-dimensional xyz space, then $\mathbb{R}^2$ becomes a subspace of $\mathbb{R}^3$.
2. For any nonnegative integers n and m with $n < m$, $F_n[x]$ is a subspace of $F_m[x]$. Also, $F_n[x]$ and $F_m[x]$ are both subspaces of $F[x]$.
3. $U_n(\mathbb{R})$ (the set of upper triangular $n \times n$ matrices with entries in $\mathbb{R}$) is a subspace of the $\mathbb{R}$–vector space $M_n(\mathbb{R})$.
4. $\mathbb{Q}[\sqrt{2}]$ is a subspace of the $\mathbb{Q}$–vector space $\mathbb{Q}[\sqrt{2}, \sqrt{3}]$. Of course, we know very well by now that since $\mathbb{Q} \subseteq \mathbb{Q}[\sqrt{2}]$, $\mathbb{Q}[\sqrt{2}]$ is directly a $\mathbb{Q}$–vector space. Both $\mathbb{Q}$–vector space structures on $\mathbb{Q}[\sqrt{2}]$ are the same, that is, in both ways of looking at $\mathbb{Q}[\sqrt{2}]$ as a $\mathbb{Q}$–vector space, the rules for vector addition and scalar multplication are the same. In the first way (viewing $\mathbb{Q}[\sqrt{2}]$ as a subspace of $\mathbb{Q}[\sqrt{2}, \sqrt{3}]$), we first think of any element $a + b\sqrt{2}$ of $\mathbb{Q}[\sqrt{2}]$ as the element $a + b\sqrt{2} + 0\sqrt{3} + 0\sqrt{6}$ of $\mathbb{Q}[\sqrt{2}, \sqrt{3}]$. Doing so, the vector sum of $a + b\sqrt{2} + 0\sqrt{3} + 0\sqrt{6}$ $(= a + b\sqrt{2})$ and $a' + b'\sqrt{2} + 0\sqrt{3} + 0\sqrt{6}$ $(= a' + b'\sqrt{2})$ is $(a + a') + (b + b')\sqrt{2} + 0\sqrt{3} + 0\sqrt{6}$ $(= (a + a') + (b + b')\sqrt{2})$. On the other hand, viewing $\mathbb{Q}[\sqrt{2}]$ directly as a $\mathbb{Q}$–vector space, the vector sum of $a + b\sqrt{2}$ and $a' + b'\sqrt{2}$ is also $(a + a') + (b + b')\sqrt{2}$. In a similar manner, you can see that the rules for scalar multiplication are also identical.
5. The example above generalizes as follows: Suppose $F \subseteq K \subseteq L$ are fields. The field extension L/F makes L an F–vector space. Since K is closed with respect to vector addition and scalar multiplication, K becomes a subspace of L. But the field extension K/F exhibits K directly as an F–vector space. The two F–vector space structures on K, one that we get from viewing K as a subspace

of the F-vector space L and the other that we get directly from the field extension K/F, are the same.

6. For any field F, $F[x^2]$ (that is, the set of all polynomials of the form $\sum_{i=0}^{n} f_i x^{2i}$, $n \geq 0$) is a subspace of $F[x]$. What is the dimension of this subspace? Can you discover a basis for this subspace? How might you generalize this example to other subspaces of $F[x]$?
7. Let V be a vector space over a field F, and let S be any nonempty subset of V. The *linear span of* S is defined as the set of all linear combinations of elements of S, that is, the set of all vectors in V that can be written as $c_1 s_1 + c_2 s_2 + \cdots + c_k s_k$ for some integer $k \geq 1$, some scalars c_i, and some vectors $s_i \in S$. Can you show that the linear span of S is a subspace of V?

   For instance, in $\mathbb{R}^3$, if we take $S = \{\mathbf{i}, \mathbf{j}\}$, then the linear span of S is the set of all vectors in $\mathbb{R}^3$ that are of the form $a\mathbf{i} + b\mathbf{j}$ for suitable scalars a and b, in other words, the xy-plane. As we saw in Example 3.23.1 above, the xy-plane is a subspace of $\mathbb{R}^3$!

We will revert to the study of field extensions in the next chapter.

Exercises

1. Deduce from the parallelogram law of addition of vectors in $\mathbb{R}^2$ that the sum of (a, b) and (a', b') is $(a + a', b + b')$.
2. Starting from the vector space axioms, prove that the properties listed in Remark 3.3 hold for all vector spaces.
3. We saw in the text on page 72 that every vector in $\mathbb{R}^2$ can be written as a linear combination of $\mathbf{i}$ and $\mathbf{w}$, where $\mathbf{w} = (1/\sqrt{2}, 1/\sqrt{2})$. This can also be seen directly, without recourse to the vector $\mathbf{j}$. Given a vector v with tip at (a, b), show that there exist scalars s and t such that $v = s\mathbf{i} + t\mathbf{w}$ by equating the x and y components of v and of the vector $s\mathbf{i} + t\mathbf{w}$, and obtaining a system of simultaneous equations for the unknowns s and t. Why does this equation have a solution?
4. Prove that the polynomials $1, 1 + x, (1 + x)^2, (1 + x)^3, \ldots$ also form a basis for $\mathbb{R}[x]$ as a $\mathbb{R}$-vector space. (Hint: To show that these poly-

nomials span $\mathbb{R}[x]$, it is sufficient to show that the polynomials $1, x, x^2, \ldots$ are in the linear span (see Example 3.23.7 above) of $1, 1+x, (1+x)^2, (1+x)^3, \ldots$ (Why?) The vector 1 is of course in the linear span. Assuming inductively that you have shown that x^{n-1} is in the linear span, show that x^n is also in the linear span by considering the binomial expansion of $(1+x)^n$. As for linear independence, suppose that $\sum_{i=0}^{n} d_i(1+x)^i = 0$. You may assume that $d_n \neq 0$ (why?) Now expand each term $(1+x)^i$ above and consider the coefficient of x^n. What do you find?)

5. Show that the matrices $e_{i,j}$ and $\sqrt{2}e_{i,j}$ $(1 \le i, j \le 2)$ form a basis for $M_2(\mathbb{Q}[\sqrt{2}])$ considered as a $\mathbb{Q}$–vector space. ($\sqrt{2}e_{i,j}$ is the 2×2 matrix with $\sqrt{2}$ in the (i,j) slot, and zeros in the remaining slots.) Now discover a basis for $M_2(\mathbb{C})$ considered as a vector space over $\mathbb{R}$.

6. Show that the set of all matrices in $M_n(\mathbb{R})$ whose trace is zero is a subspace of $M_n(\mathbb{R})$. (Recall that the *trace* of a matrix is simply the sum of the diagonal entries.) Discover a basis for this subspace, if you are told that the dimension of this subspace is $n^2 - 1$.

7. Let S be the subset of $\mathbb{R}^\infty$ consisting of all vectors $(a_0, a_1, \ldots)$ in which only finitely many of the a_i are nonzero. Show that S is a subspace of $\mathbb{R}^\infty$. Is S finite dimensional? Discover a basis for S.

8. This exercise is designed to allow you to deduce two corollaries to some of the results that we have developed in this chapter. Here is a hint that applies to both parts: try to combine the statement of various results we have arrived at in the chapter with the statement of Corollary 3.17!

   (a) Let V be an n-dimensional vector space over a field F. Suppose $S = \{v_1, v_2, \ldots, v_n\}$ is a set of n vectors in V that spans V. Prove that S is a basis for V.

   (b) Let V be an n-dimensional vector space over a field F. Suppose $B = \{v_1, v_2, \ldots, v_n\}$ is a set of n vectors in V that are linearly independent. Prove that B is a basis for V.

9. If V is a finite-dimensional vector space and if W is a subspace of V, prove that the dimension of W is no bigger than the dimension of V. Now prove that if the dimension of W and V are equal, then

$W = V$. (Hint: Notice that any linearly independent subset of W is also a linearly independent subset of V. Now look to Theorem 3.19 and Exercise 8b above for inspiration!)

10. Show that the *nth Bernstein Poylnomials* $B_i^n(x) = \binom{n}{i}x^i(1-x)^{n-i}$, $(i = 0, 1, \ldots, n)$ form a basis for $\mathbb{R}_n[x]$ $(n \geq 1)$ as follows:
    (a) Show that $1 = \sum_{i=0}^{n} B_i^n$.
    (b) The equation in part 10a above continues to hold if we replace n by $n-1$ everywhere. (Why?) Make this replacement, multiply throughout by x, and derive the relation $x = \sum_{i=0}^{n}(i/n)B_i^n$. (Hint: you will need to use the relation $\binom{n-1}{i-1} = (i/n)\binom{n}{i}$. Why does this last relation hold?)
    (c) Similarly, for $k = 2, \ldots, n-1$, show that $x^k = \sum_{i=0}^{n}(i(i-1)\cdots(i-k+1)/n(n-1)\cdots(n-k+1))B_i^n$.
    (d) Now conclude that the B_i^n span $\mathbb{R}_n[x]$.
    (e) Use Exercise 8a above to conclude that the B_i form a basis.

    These Bernstein polynomials find applications in diverse areas of mathematics, as well as in various applied fields, such as computer graphics! For instance, in advanced calculus, they are useful in showing that any continuous function on an interval $[a, b]$ can be approximated arbitrarily closely by a *polynomial* function. (This is known as the *Weierstrass Approximation Theorem.*) In computer graphics, they are used to fit, through a given set of points, a curve that is smooth and has minimal "wiggle,"and as well, to provide convenient handles by which the user can then control the shape of this curve.

11. Let V be a 2-dimensional vector space over a field F. Let $\{v_1, v_2\}$ be a basis for V. A *linear transform* on V is a function $T\colon V \to V$ that satisfies the properties (1) $T(u + v) = T(u) + T(v)$ and (2) $T(au) = aT(u)$ for all vectors u and v and for all scalars a. Let T be a linear transform on V.
    (a) Show that there exist *unique* scalars a, b, c, and d such that $T(v_1) = av_1 + bv_2$, and $T(v_2) = cv_1 + dv_2$.
    (b) Show that if an arbitrary vector u in V is expressed as $xv_1 + yv_2$ for suitable scalars x and y (why is this possible?), then $T(u) = (ax + cy)v_1 + (bx + dy)v_2$. (Hint: Recall what it means for T to be a linear transform.)

(c) In an expression such as $u = xv_1 + yv_2$, think of the scalar x as the "coordinate" of u along the "axis" represented by the basis vector v_1, and think of the scalar y as the "coordinate" of u along the "axis" represented by the basis vector v_2. Let $T(u)_1$ and $T(u)_2$ denote the coordinates of the vector $T(u)$ along the v_1 axis and along the v_2 axis. Show that we have the matrix equation

$$\begin{pmatrix} T(u)_1 \\ T(u)_2 \end{pmatrix} = \begin{pmatrix} a & c \\ b & d \end{pmatrix} \begin{pmatrix} x \\ y \end{pmatrix}.$$

(Note that the 2×2 matrix above is *not* $\begin{pmatrix} a & b \\ c & d \end{pmatrix}$ as one might naively expect!)

(d) Conversely, given a matrix $\begin{pmatrix} a & c \\ b & d \end{pmatrix}$ of scalars, check that the function $T: V \to V$ that for any pair of coordinates x and y sends the vector $u = xv_1 + yv_2$ to the vector $(ax + cy)v_1 + (bx + dy)v_2$ is a linear transform on V. (Hint: Let $u' = x'v_1 + y'v_2$ be another vector. Check that $T(u + u') = T(u) + T(u')$. Similarly, if r is any scalar, check that $T(ru) = rT(u)$.)

We thus obtain a correspondence between linear transforms on V and elements of $M_2(F)$, and this correspondence can be checked to be one-to-one and onto. (Note that this correspondence was derived by using a fixed basis, $\{v_1, v_2\}$.)

The results of this exercise generalize to correspondences between linear transforms on n-dimensional vector spaces and elements of $M_n(F)$ as well. You might want to play with the case $n = 3$.

Notes

Remarks on Example 3.2.4 It is worth remarking that our definition of scalar multiplication is a very natural one. First, observe that we can consider $\mathbb{R}$ to be a subring of $M_n(\mathbb{R})$ in the following way: the set of diagonal matrices of the form diag(r), as r ranges through $\mathbb{R}$, is essentially the same as $\mathbb{R}$ (see the notes to Example 2.12.8 on page 61). (Observe that this makes the set of diagonal matrices of the form diag(r) a field in its own right!) Under this identification of $r \in \mathbb{R}$ with $\mathrm{diag}(r)$, what is the most natural way to multiply a scalar r and a vector $((a_{i,j}))$? Well, we think of r as $\mathrm{diag}(r)$, and then define $r \cdot ((a_{i,j}))$ as just the usual product of the two matrices $\mathrm{diag}(r)$ and $((a_{i,j}))$. But, as you can check easily, the product of $\mathrm{diag}(r)$ and $((a_{i,j}))$ is just $((ra_{i,j}))$! It is in this sense that our definition of scalar multiplication is natural—it arises from the rules of matrix multiplication itself. Notice that once $\mathbb{R}$ has been identified with the subring of $M_n(\mathbb{R})$ consisting of the set of diagonal matrices of the form $\mathrm{diag}(r)$, this example is just another special case of Example 3.2.7.

Remarks on Example 3.2.10 $(V, +)$ remains an abelian group. This does not change when we restrict our attention to the subfield F. So we only need to worry about what the new scalar multiplication ought to be. But there is a natural way to multiply any element f of F with any element v of V: simply consider f as an element of K, and use the multiplication already defined between elements of K and elements of V! The scalar multiplication axioms clearly hold: for any f and g in F and any v and w in V, we may first think of f and g as elements of K, and since the scalar multiplication axioms hold for V viewed as a vector space over K, we certainly have $f \cdot (v + w) = f \cdot v + f \cdot w$, $(f + g) \cdot v = f \cdot v + g \cdot v$, $(fg) \cdot v = f \cdot (g \cdot v)$, and $1 \cdot v = v$.

Remarks on Example 3.11.9 This example is a bit tricky. Why are the e_i not a basis? They are certainly linearly independent, since if $\sum_{i=0}^{n} c_i e_i = 0$ for some scalars $c_i \in F$, then the tuple $(c_0, c_1, \ldots, c_n, 0, 0, \ldots)$ must be zero, but a tuple is zero if and only if each of its components is zero. Thus, each of c_0, c_1, ..., c_n must be zero, proving linear independence. However, the e_i *do not span* F^∞, contrary to what one might expect. To understand this, let us look at something that has been implicit all along in the definition of linear combination. The e_i would span F^∞ if every vector in F^∞ could be written as a linear combination of elements of the set $\{e_0, e_1, e_2, \ldots\}$. Now notice that whenever we consider linear combinations, we only consider sums of a *finite* number of terms. Hence, a linear combination of elements of the set $\{e_0, e_1, e_2, \ldots\}$ looks like $c_{i_1} e_{i_1} + c_{i_2} e_{i_2} + \cdots + c_{i_n} e_{i_n}$ for some *finite* n. It is clear that any vector that is expressible in such a manner will have only finitely many components that are nonzero. (These will be at most the ones at the slots i_1, i_2, ..., i_n; all other components will be zero.) Consequently, the vectors in

F^∞ in which infinitely many components are nonzero (for example, the vector $(1, 1, 1, \ldots)$), cannot be expressed as linear combinations of the e_i. See Exercise 7.

It is worth pointing out that infinite sums have no algebraic meaning. Addition is, to begin with, a binary operation, that is, it is a rule that assigns to a_1 and a_2 the element $a_1 + a_2$. This can be extended inductively to a finite number of a_i: for instance, the sum $a_1 + a_2 + a_3 + a_4 + a_5$ is defined as $(((a_1 + a_2) + a_3) + a_4) + a_5$. (In other words, we first determine $a_1 + a_2$, then we add a_3 to this, then a_4 to what we get from adding a_3, and then finally a_5 to what we got at the previous step.) While this inductive definition makes sense for a *finite* number of terms, it makes no sense for an infinite number of terms. To interpret infinite sums of elements, we really need to have a notion of convergence (such as the ones you may have seen in a course on real analysis). Such notions may not exist for arbitrary fields.

Remarks on Zorn's Lemma As mentioned in the text, Zorn's "Lemma" is really not a lemma; it is a fundamental axiom of logic. It is equivalent to one other axiom that you may have seen, the Axiom of Choice. (As well, these two axioms are both equivalent to several other axioms—see any introductory textbook on logic!)

So what is Zorn's Lemma all about? What follows is a *loose* description. We necessarily have to suppress many details, since to provide a full treatment of Zorn's Lemma, we need to develop several concepts, and this would carry us too far afield. As always, however, you are encouraged to read more about this topic on your own.

As you may guess from the context in which we invoked it in Theorem 3.14, Zorn's Lemma guarantees the existence of "maximal" elements in certain kinds of sets. What does this mean? First, we need a set S, and on this set, we need a reasonable notion of one element being larger than another. What is a "reasonable" notion is a well-understood matter: there are definite properties that a reasonable notion ought to satisfy (which we will not develop here). A notion of one element being larger than another that satisfies these properties is known as an *order* on the set. Given such a reasonable notion (or order) on S, if it has the further property that given any two elements of S, one can always determine whether one is larger than the other, we call this order a *total* or *linear* order. (For instance, the usual notion of one integer being larger than another is a total order on the set of integers.) The interesting situation that is relevant to proving that every vector space has a basis is that in which we have only a *partial* order on S. This situation occurs when it is possible to tell whether one element is bigger than another only for *certain* pairs of elements in S. (For instance, in the set of all positive integers, we may *decree* m to be less than or equal to n if and only if m divides n. This is a partial order, but this does not allow us to compare two positive integers m and n if neither divides the other, and is therefore not a total order.) Finally, given an order on S, an element x of S is said to be maximal with respect to this order if for any other element y of S, either x is bigger than y or else x and y cannot be compared. (It must be kept in mind that maximal elements need not be unique. For instance, in the set of all integers between 1 and 12, if we were to define $m \leq n$ if $m|n$, then 12 is clearly maximal, but so are 11, 10, 9, 8, and 7—why?)

Given a partial order on S, the following question arises: Does S have a maximal element? In general, it need not, but Zorn's Lemma guarantees that if the partial order has one more property (which we will not state here), then S will have a maximal element.

In the proof of Theorem 3.14, the set to which we apply Zorn's Lemma is the set S of all linearly independent subsets of V. (Thus, each element of S is itself a set, consisting of linearly independent vectors of V.) Given two elements S_1 and S_2 of S, we define $S_1 \leq S_2$ if and only if $S_1 \subseteq S_2$. (Remember, both S_1 and S_2 are themselves sets!) With this notion, one can check that we have an order on S. This order is only a partial order, since it is possible to find two subsets of linearly independent vectors of V, neither of which contains the other. (For instance, if v and w are two distinct nonzero vectors and if $S_1 = \{v\}$ and $S_2 = \{w\}$, then S_1 and S_2 are both in S, but S_1 and S_2 cannot be compared, since neither $S_1 \subseteq S_2$ nor $S_2 \subseteq S_1$.) But what is relevant is the following: the set S with this partial order satisfies the extra hypothesis in Zorn's Lemma, so we are guaranteed that there is at least one maximal element in S. Let us pick any one maximal element and call it T. If y is any vector in V, then $T' = T \cup \{y\}$ must be a linearly dependent set; otherwise T' would be a linear independent set that strictly contains T (or, in the language of our order, T' would be bigger than T), violating the maximality of T in the set S. As we saw in the proof of Theorem 3.14, T must therefore be a basis of V.

Remarks on Theorem 3.20, the general case The proof of this theorem when V is not assumed to be finite-dimensional involves just a minor modification of the proof of Theorem 3.14. What we need to show is that there is a maximal linearly independent subset B of V *that contains* C. Then, exactly as in the proof of Theorem 3.14, this maximal linearly independent set would be a basis of V, and of course, it would have been chosen so as to contain C. To show the existence of B, we need to consider the set S of all linearly independent subsets of V *that contain* C. One would impose a partial order on this set exactly as before (see the remarks above on the proof of Theorem 3.14). Once again, S, with this partial order, will turn out to satisfy the extra hypothesis of Zorn's Lemma, and will hence have a maximal element. That maximal element would be our desired maximal linearly independent subset of V that contains C.

4 CHAPTER Field Extensions

Recall that at the beginning of the previous chapter, we were interested in the following problem: Given a field extension K/F, how does one measure how big K is relative to F? In fact, the whole chapter on vector spaces was introduced primarily to solve this problem. With the material on vector spaces under our belt, we can now give our answer.

Recall from Example 3.2.6 that K has the structure of an F-vector space. It turns out that the size of K/F as a field extension is best measured by the dimension of K as an F-vector space!

We are now ready to consider field extensions in greater depth. But first, since the dimension of K as an F-vector space plays such an important role in the study of field extensions, we give it a special name:

Definition 4.1
Given a field extension K/F, the dimension of K as an F-vector space is called the *degree* of K over F, and is denoted $[K : F]$.

One of the first questions that we will tackle arises in the following situation: Suppose that F, K, and L are three fields with $F \subseteq K \subseteq L$. (For instance, take $F = \mathbb{Q}$, $K = \mathbb{Q}[\sqrt{2}]$, and $L = \mathbb{Q}[\sqrt{2}, \sqrt{3}]$. We know by now that both $\mathbb{Q}[\sqrt{2}]$ and $\mathbb{Q}[\sqrt{2}, \sqrt{3}]$ are actually fields.

We have used this fact several times already in the previous chapter.) These three fields give us three field extensions—K/F, L/K, and L/F. A natural question is the following: How are the three degrees $[K : F]$, $[L : K]$, and $[L : F]$ related to one another? It turns out that there is a very simple relationship, which we will describe in the theorem below.

Theorem 4.2

Suppose that $F \subseteq K \subseteq L$ are fields.

1. *If $[L : F]$ is finite, then $[L : K]$ and $[K : F]$ are also finite.*
2. *Suppose that $[L : K]$ and $[K : F]$ are finite, with $[L : K] = m$ and $[K : F] = n$. If $\mathcal{B} = \{v_1, \ldots, v_m\}$ is a basis for L as a K-vector space, and if $\mathcal{C} = \{w_1, \ldots, w_n\}$ is a basis for K as an F-vector space, then $\mathcal{A} = \{v_i w_j \mid i = 1, \ldots, m,\ j = 1, \ldots, n\}$ is a basis for L as an F-vector space. In particular, $[L : F]$ is also finite, and $[L : F] = [L : K] \cdot [K : F]$.*

Proof The role of K in this proof is rather delicate, and it is worth paying some attention to it. On the one hand, the field extension L/K makes L a K-vector space, and K thus acts as *scalars* for this vector space. On the other hand, the field extension K/F makes K an F-vector space, so in this context, K acts as a set of *vectors*. Thus, K *simultaneously* plays the role of scalars (when L is considered as a K-vector space), and vectors (when K is considered as an F-vector space). No confusion should arise if the context in which K appears is always kept in mind.

In general, you will find it easier to keep track of all the arguments in the proof if you pay careful attention to what the scalars are and what the vectors are whenever a particular vector space is being considered.

Now, just to keep yourselves on your toes, note that the F-vector space structure on K is the same as that obtained by considering K as a subspace of the F-vector space L! (See Examples 3.23.4 and 3.23.5.)

(If all this already seems rather bewildering, read the notes on page 115 before going any further.)

Let us first prove part 1. To prove that $[K : F]$ is finite, recall the statement of Exercise 9 of Chapter 3: If V is a finite-dimensional vector space and if W is a subspace of V, then the dimension of W

is no bigger than the dimension of V. In particular, W must also be finite dimensional. Now apply this to our situation: K is a subspace of the F-vector space L, and we are given that $[L : F]$ is finite. It follows that $[K : F]$ must be finite!

As for the finiteness of $[L : K]$, suppose $\{z_1, \ldots, z_t\}$ (where $t = [L : F]$) is some basis for L as an F-vector space. This means that every element in L can be written as a linear combination $\sum_{i=1}^{t} f_i z_i$ for suitable elements $f_i \in F$. Since $F \subseteq K$, *the f_i can also be considered as elements of K!* This means that every element in L can be written as a K-linear combination of the elements $\{z_1, \ldots, z_t\}$, that is, the elements $\{z_1, \ldots, z_t\}$ span L as *a K-vector space.* By Theorem 3.12, some subset of $\{z_1, \ldots, z_t\}$ must be a basis for L considered as a K-vector space. It follows that $[L : K] \leq t$, that is, $[L : K]$ must be finite.

Now for part 2. Assume for the moment that we have already proved that $\mathcal{A}$ is a basis for L as an F-vector space. We will see how the claim $[L : F] = [L : K][K : F]$ follows from this. Since we are assuming that $\mathcal{A}$ is a basis for L as an F-vector space, $[L : F]$ must equal the number of elements in $\mathcal{A}$. Given that $\mathcal{B}$ has m elements and $\mathcal{C}$ has n elements, how many elements should $\mathcal{A}$ have then? One is tempted to say m times n. This is indeed correct, but something needs to be verified first. Given two different pairs of indices (i, j) and (i', j') ("different" means that either $i \neq i'$ or $j \neq j'$), what if $v_i w_j$ were to equal $v_{i'} w_{j'}$? If this were to happen, the number of elements in $\mathcal{A}$ would be *less than mn* because of the duplication. Let us show that this cannot happen. We have two cases to consider, either $i = i'$ or $i \neq i'$. If $i = i'$, then our relation would read $v_i(w_j - w_{j'}) = 0$. View this as the product of the vector v_i and the scalar $(w_j - w_{j'})$ in the K-vector space L. Since v_i is a basis element for L as a K-vector space, it is not zero, so $w_j - w_{j'}$ must be zero. Thus, $w_j = w_{j'}$. But the elements of $\mathcal{C}$ (being basis elements for K as an F-vector space) are all distinct, so the only way w_j could equal $w_{j'}$ is if $j = j'$. But this says that the pair (i, j) and (i', j') are the same, which contradicts our assumption. As for the second case $(i \neq i')$, once again think of w_j and $w_{j'}$, which are elements of K, as scalars for the K-vector space L. Since v_i and $v_{i'}$ are two distinct basis elements for L considered as a K-vector space, the equation $w_j v_i + (-w_{j'}) v_{i'} = 0$ implies that the scalars w_j and $w_{j'}$ must both be zero. But w_j and $w_{j'}$, being members of $\mathcal{C}$, are basis elements of K considered as an F–vector space, and are hence not zero. This is a contradicition! Thus, given two different pairs of indices (i, j) and (i', j'), $v_i w_j$ could never equal $v_{i'} w_{j'}$. It follows that $\mathcal{A}$ must have precisely mn elements, so $[L : F]$ indeed equals $[L : K][K : F]$ (and of course, $[L : F]$ is indeed finite).

We still have the task of proving that $\mathcal{A}$ is a basis for L as an F–vector space!

Let us first show that the elements of $\mathcal{A}$ span L as an F–vector space. That is, we need to show that every element of L can be written as an F–linear combination of elements of $\mathcal{A}$. Consider an arbitrary element x of L. Viewing x as a vector in the K–vector space L, x may be written as a linear combination of vectors from $\mathcal{B}$ (since $\mathcal{B}$ is a basis for the K–vector space L). Hence, there exist elements $k_1, \ldots, k_m$ in K (with the k_i to be viewed as *scalars*) such that

$$x = k_1 \cdot v_1 + \cdots + k_m \cdot v_m. \tag{4.1}$$

Now view each of the k_i above as *vectors* of the F–vector space K. Since $\mathcal{C}$ is a basis for this vector space, for each i there exist elements $f_{i,j} \in F$ $(j = 1, \ldots, n)$, with the $f_{i,j}$ to be viewed as scalars, such that

$$k_i = f_{i,1} \cdot w_1 + \cdots + f_{i,n} \cdot w_n. \tag{4.2}$$

Now temporarily view Equations 4.1 and 4.2 not as equations concerning vectors and scalars, but as equations concerning elements of the field L. We may then plug the expression on the right-hand side of Equation 4.2 for the k_i into the right-hand side of Equation 4.2, and we find that

$$\begin{aligned} x = {} & (f_{1,1} \cdot w_1 + \cdots + f_{1,n} \cdot w_n) \cdot v_1 \\ & + \cdots + \\ & (f_{m,1} \cdot w_1 + \cdots + f_{m,n} \cdot w_n) \cdot v_m. \end{aligned} \tag{4.3}$$

We may rewrite Equation 4.3 as

$$\begin{aligned} x = {} & f_{1,1} \cdot (v_1 w_1) + \cdots + f_{1,n} \cdot (v_1 w_n) \\ & + \cdots + \\ & f_{m,1} \cdot (v_m w_1) + \cdots + f_{m,n} \cdot (v_m w_n). \end{aligned} \tag{4.4}$$

But this just says that every element in L can be written as an F-linear combination of the elements $v_i w_j$. Viewing the $v_i w_j$ as *vectors* in the F-vector space L, we find that the $v_i w_j$ span L.

Now we need to show that the elements of $\mathcal{A}$ are linearly independent. That is, if

$$\begin{aligned} f_{1,1}(v_1 w_1) + \cdots + f_{1,n}(v_1 w_n) \\ + \cdots + \\ f_{m,1}(v_m w_1) + \cdots + f_{m,n}(v_m w_n) = 0 \end{aligned} \tag{4.5}$$

for some scalars $f_{i,j}$ ($i = 1, \ldots, m$, $j = 1, \ldots, n$), then we must show that each of the scalars $f_{i,j}$ must be zero. For this, view Equation 4.5 as an equation concerning elements of the field L, and group all the v_i terms to obtain

$$\begin{aligned} (f_{1,1} w_1 + \cdots + f_{1,n} w_n) v_1 \\ + \cdots + \\ (f_{m,1} w_1 + \cdots + f_{m,n} w_n) v_m = 0. \end{aligned} \tag{4.6}$$

Let

$$\begin{aligned} k_1 &= f_{1,1} w_1 + \cdots + f_{1,n} w_n, \\ &\vdots \qquad\qquad \vdots \\ k_m &= f_{m,1} w_1 + \cdots + f_{m,n} w_n. \end{aligned} \tag{4.7}$$

Where do the k_i live? The $f_{i,j}$ are all elements of F, while the w_j are all elements of K, so each product $f_{i,j} w_j$ is an element of K. Since each k_i is just a sum of products of the form $f_{i,j} w_j$, we find that each k_i is an element of K. Now rewrite Equation 4.6 as

$$k_1 v_1 + \cdots + k_m v_m = 0, \tag{4.8}$$

and view this as an equation concerning the *vectors* v_i and the *scalars* k_i of the K-vector space L. Because the v_i are all K-linearly independent (as $\mathcal{B}$ is a basis for the K-vector space L), we find that each k_i must be zero. Now plug 0 for each k_i in 4.7 to obtain the m equations

$$\begin{aligned} 0 &= f_{1,1} w_1 + \cdots + f_{1,n} w_n, \\ &\vdots \qquad\qquad \vdots \\ 0 &= f_{m,1} w_1 + \cdots + f_{m,n} w_n. \end{aligned} \tag{4.9}$$

View each of these m equations as equations concerning the *vectors* w_j and the *scalars* $f_{i,j}$ of the F-vector space K. By the F-linear independence of the w_j, the first of the m equations of 4.9 shows that $f_{1,1} = \ldots = f_{1,n} = 0$, the second of the m equations shows that $f_{2,1} = \ldots = f_{2,n} = 0, \ldots$, and the last of the m equations of 4.9 shows that $f_{m,1} = \ldots = f_{m,n} = 0$. Hence all the $f_{i,j}$ are zero, establishing the F-linear independence of $\mathcal{A}$. This proves that $\mathcal{A}$ is a basis for L as an F-vector space, and establishes the theorem. □

Remark 4.3
Putting together parts 1 and 2 of this theorem, we find that if $F \subseteq K \subseteq L$ are fields with $[L : F]$ finite, then $[L : F] = [L : K][K : F]$.

Example 4.4
Let us expand on the example described before the theorem: Consider $F = \mathbb{Q}$, $K = \mathbb{Q}[\sqrt{2}]$, and $L = \mathbb{Q}[\sqrt{2}, \sqrt{3}]$. The set $\{1, \sqrt{3}\}$ forms a K-basis for L—recall from Definition 3.10 that this just means a basis for L viewed as a K-vector space; we use the term "K-basis" to emphasize the field of scalars. (By Example 2.15.6 and Example 2.12.4, every element of L can be written as $a + b\sqrt{2} + c\sqrt{3} + d\sqrt{6}$ for suitable a, b, c, and d in $\mathbb{Q}$. We rewrite this expression as $(a + b\sqrt{2}) + (c + d\sqrt{2})\sqrt{3}$. Since the expressions in the parentheses are elements of K, 1 and $\sqrt{3}$ span L as a K-vector space. Can you now use the results of Exercise 7b of Chapter 2 to show that 1 and $\sqrt{3}$ are also K-linearly independent?) Thus, we may take $\mathcal{B} = \{1, \sqrt{3}\}$. Similarly, the set $\{1, \sqrt{2}\}$ forms a $\mathbb{Q}$-basis for K. (We've seen this before!) Thus, we may take $\mathcal{C} = \{1, \sqrt{2}\}$. According to the theorem, a $\mathbb{Q}$-basis for L should be given by the products of elements from $\mathcal{B}$ and $\mathcal{C}$, that is, by the set $\{1 \cdot 1, 1 \cdot \sqrt{2}, \sqrt{3} \cdot 1, \sqrt{3} \cdot \sqrt{2}\}$, which is just $\{1, \sqrt{2}, \sqrt{3}, \sqrt{6}\}$. Can we verify this independently? Of course! We have already done so! Exercise 7d of Chapter 2 established the $\mathbb{Q}$-linear independence, and the fact that $\{1, \sqrt{2}, \sqrt{3}, \sqrt{6}\}$ spans L as a $\mathbb{Q}$-vector space is clear.

Let us turn to another question of fundamental interest in the study of field extensions. Let K/F be a field extension, and let a be an arbitrary element of K. We have already seen the notion of the subring of K generated by F and a in Chapter 2. (Remember, every field is first of all a ring, so it makes sense to talk of subrings of K

generated by F and other elements.) We saw (Lemma 2.14) that the set $F[a]$ of all polynomial expressions in a with coefficients in F is a subring of K. In fact, we saw in the discussion preceeding Lemma 2.14 that $F[a]$ is the smallest subring of K containing F and a, in the sense that any other subring of K that contains F and a must also contain $F[a]$.

Now here is the question: Since in this particular situation, K and F are not merely commutative rings but actually fields, is there a corresponding notion of *the subfield of K generated by F and a?* What could such a notion possibly mean? The sub*ring* of K generated by F and a was just the smallest subring of K containing F and a (in the sense described above). The sub*field* of K generated by F and a should correspondingly be the smallest subfield of K containing F and a—where by "smallest" we mean that any other subfield of K containing F and a must also contain this subfield.

But can we find such a smallest subfield containing F and a? In other words, is there a subset of K that has the following three properties: First, it contains both F and a; second, it is a field in its own right; and third, it is contained in every other subfield of K that contains both F and a? The answer is yes! In fact, it is not hard to describe this subset.

Assuming that such a subset S exists, let us try to discover what it must look like. (Once we have proved that S exists, we will switch to a more suggestive notation for this subset.) First of all, since S contains a and must be closed under multiplication, it must contain $a^2, a^3, \ldots$ Next, since S contains F, it must also contain all products of the form fa^i, where $f \in F$ and $i \geq 0$. Once it contains products of the form fa^i, it must contain all *sums* of such products, that is, it must contain all expressions of the form $f_0 + f_1a + f_2a^2 + \cdots + f_na^n$, where $n \geq 0$, and the $f_i \in F$.

What we have shown is that S must contain $F[a]$! We could of course have seen this directly from our requirement that S be a subfield of K containing F and a—every field is also a ring, so S is also a sub*ring* of K containing F and a, and since $F[a]$ is the smallest subring of K containing F and a, S must contain $F[a]$!

We are not done yet with the task of describing S. Now let us now invoke our requirement that S be a field. We have already seen that every expression of the form $f_0 + f_1a + f_2a^2 + \cdots + f_na^n$ that is nonzero is in S, but because S must be a field, the *multiplicative inverse* of this element (which exists in K because K is a field and contains multiplicative inverses of all its nonzero elements) must also be in S. In other words, S must contain $1/(f_0 + f_1 a + f_2 a^2 + \cdots + f_n a^n)$. From this it follows (why?) that S must contain all products of expressions of the form $f_0 + f_1 a + f_2 a^2 + \cdots + f_m a^m$, ($m \geq 0$, $f_i \in F$), and expressions of the form $1/(g_0 + g_1 a + g_2 a^2 + \cdots + g_n a^n)$ ($n \geq 0$, $g_j \in F$, $g_0 + g_1 a + g_2 a^2 + \cdots + g_n a^n \neq 0$), that is, S must contain all quotients of the form $\dfrac{f_0 + f_1 a + f_2 a^2 + \cdots + f_m a^m}{g_0 + g_1 a + g_2 a^2 + \cdots + g_n a^n}$. Is that enough? Will that give us our "smallest" subfield of K containing F and a? Yes! This is the content of the following theorem:

Theorem 4.5

Let K/F be a field extension, and let a be any element of K. Let $F(a)$ denote the set

$$\left\{ \frac{f_0 + f_1 a + f_2 a^2 + \cdots + f_m a^m}{g_0 + g_1 a + g_2 a^2 + \cdots + g_n a^n} \right\}$$

($m \geq 0$, $n \geq 0$, $f_i, g_j \in F$, $g_0 + g_1 x a + g_2 a^2 + \cdots + g_n a^n \neq 0$). Then $F(a)$ is a subfield of K that contains both F and a. Moreover, it is the smallest subfield of K that contains both F and a, that is, if E is any other subfield of K that contains both F and a, then $F(a) \subseteq E$.

Remark 4.6

$F(a)$ is known as the *subfield of K generated by F and a*. Note the parentheses around a that are used to distinguish the subfield of K generated by F and a from $F[a]$. Since every polynomial expression $f_0 + f_1 a + f_2 a^2 + \cdots + f_m a^m$ can be written as $f_0 + f_1 a + f_2 a^2 + \cdots + f_m a^m / 1 + 0a + 0a^2 + \cdots$, it is clear that $F[a] \subseteq F(a)$.

Proof (Sketch:) Use Lemma 2.11 to show that $F(a)$ is a sub*ring* of K. (Thus, you need to show that $F(a)$ is closed with respect to addition and multiplication, that $1 \in F(a)$, and that for all $s \in F(a)$, $-s \in F(a)$ also.) Since $F(a)$ is contained in K, $F(a)$ is commutative and has no zero-divisors. (Explain!) To see that $F(a)$ is a field, let x be a nonzero element of $F(a)$. We need to show that x has a multiplicative inverse in $F(a)$. (Note that x has a multiplicative inverse in K, since K is a field. The issue is whether this multiplicative inverse lives inside

$F(a)$ or not.) Write

$$x = \frac{f_0 + f_1 a + f_2 a^2 + \cdots + f_m a^m}{g_0 + g_1 a + g_2 a^2 + \cdots + g_n a^n},$$

and notice that since $x \neq 0$, $f_0 + f_1 a + f_2 a^2 + \cdots + f_m a^m \neq 0$. It is now easy to exhibit the multiplicative inverse of x—it is just the reciprocal

$$\frac{g_0 + g_1 a + g_2 a^2 + \cdots + g_n a^n}{f_0 + f_1 a + f_2 a^2 + \cdots + f_m a^m}.$$

From the very form of this inverse (it is the quotient of two polynomial expressions in a with coefficients in F), it is clear that this inverse is an element of $F(a)$.

For the last statement of the theorem, the discussion in the paragraph preceeding the theorem shows that if E is a subfield of K that contains both F and a, then, since E is closed with respect to multiplication and addition, and must contain the multiplicative inverses of all its nonzero elements, E must necessarily contain all expressions of the form

$$\frac{f_0 + f_1 a + f_2 a^2 + \cdots + f_m a^m}{g_0 + g_1 a + g_2 a^2 + \cdots + g_n a^n}$$

($m \geq 0$, $n \geq 0$, $f_i, g_j \in F$, $g_0 + g_1 a + g_2 a^2 + \cdots + g_n a^n \neq 0$), that is, E must contain $F(a)$. □

What does $F(a)$ reduce to when $a = 0$? More generally, what does $F(a)$ reduce to when $a \in F$?

The proof of the theorem leads naturally to another question. We have seen that $F[a] \subseteq F(a)$. *Can $F[a]$ possibly equal $F(a)$?* On the surface of things, the answer seems to be no. $F(a)$ seems to be larger than $F[a]$, since $F[a]$ only contains expressions of the form $f_0 + f_1 a + \cdots + f_m a^m$, while $F(a)$ contains *quotients* of such expressions. But this can be misleading. Consider the following:

Example 4.7

Let $F = \mathbb{Q}$, $K = \mathbb{R}$, and let $a = \sqrt{2}$. Let us compare $\mathbb{Q}[\sqrt{2}]$ and $\mathbb{Q}(\sqrt{2})$. We are very familiar with the fact that $\mathbb{Q}[\sqrt{2}] = \{a + b\sqrt{2} \mid a, b \in \mathbb{Q}\}$ (Example 2.15.1). What about $\mathbb{Q}(\sqrt{2})$? By a similar argument as in Example 2.15.1, $\mathbb{Q}(\sqrt{2}) = \left\{ \dfrac{a + b\sqrt{2}}{c + d\sqrt{2}} \mid c + d\sqrt{2} \neq 0 \right\}$. Of course, $\mathbb{Q}[\sqrt{2}] \subseteq \mathbb{Q}(\sqrt{2})$, but we actually have the reverse inclusion as well. For (and we have seen this too!), we can write $\dfrac{1}{c + d\sqrt{2}}$ upon rationalizing as $\dfrac{c}{c^2 - 2d^2} + \sqrt{2}\dfrac{-d}{c^2 - 2d^2}$. Hence, every expression of the form $\dfrac{a + b\sqrt{2}}{c + d\sqrt{2}}$ can be written as the product of $a + b\sqrt{2}$ and $\dfrac{c}{c^2 - 2d^2} + \sqrt{2}\dfrac{-d}{c^2 - 2d^2}$, and this product is an expression of the form $x + y\sqrt{2}$ with x and y rational numbers. (What are x and y in terms of a, b, c, and d?) This shows that $\mathbb{Q}(\sqrt{2}) \subseteq \mathbb{Q}[\sqrt{2}]$. It follows that $\mathbb{Q}[\sqrt{2}] = \mathbb{Q}(\sqrt{2})$!

Now that we have an example where $F[a] = F(a)$, we of course wonder: Must $F[a]$ *always* equal $F(a)$? The answer is no! Consider the following example:

Example 4.8

Take $F = \mathbb{R}$, and take K to be the field $\mathbb{R}(x)$ (see Example 2.23.5). Take $a = x^2$. What is $\mathbb{R}[x^2]$? This is just the set of all polynomials of the form $\{f_0 + f_1x^2 + f_2x^4 + f_3x^6 + \cdots + f_nx^{2n}\}$ $(n \geq 0, f_i \in \mathbb{R})$. Thus, $\mathbb{R}[x^2]$ just consists of all polynomials in x^2. What about $\mathbb{R}(x^2)$? Well, $\mathbb{R}(x^2)$ consists of all quotients of polynomials in x^2, that is

$$\mathbb{R}(x^2) = \left\{ \frac{f_0 + f_1x^2 + f_2x^4 + \cdots + f_mx^{2m}}{g_0 + g_1x^2 + g_2x^4 + \cdots + g_nx^{2n}} \right\}$$

(where, of course, the polynomials in the denominator are not zero). As usual, $\mathbb{R}[x^2] \subseteq \mathbb{R}(x^2)$, but we claim that that the reverse inclusion does not hold, that is, we claim that $\mathbb{R}(x^2)$ contains elements that are *not* in $\mathbb{R}[x^2]$. To see this, we do not have to look very far: consider $1/x^2 \in \mathbb{R}(x^2)$. We claim $1/x^2 \notin \mathbb{R}[x^2]$. This should be intuitively clear, but here is a formal proof: Suppose that $1/x^2 \in \mathbb{R}[x^2]$. Then $1/x^2 = c_0 + c_1x^2 + \cdots + c_nx^{2n}$ for some some $n \geq 0$ and $c_i \in \mathbb{R}$. We may assume that $c_n \neq 0$ (why?), so the polynomial $c_0 + c_1x^2 + \cdots + c_nx^{2n}$ has degree $2n$. Mutlplying through by x^2, we find $1 = c_0x^2 + c_1x^4 + \cdots + c_nx^{2n+2}$. The left-hand side is a constant, that is, a polynomial of degree 0, while the right hand side is a polynomial of degree $2n + 2$, which is a contradiction (see Remark 2.16). Hence, in this example, $F[a] \neq F(a)$.

Remark 4.9

To say that $F[a] = F(a)$ is to say that every quotient of polynomial expressions in a can be *rewritten* as a polynomial expression in a. In particular, since the multiplicative inverse of any (nonzero) polynomial expression $\sum_{i=0}^{n} f_i a^i$ is just the quotient of the polynomial expression $1\ (= 1 + 0a + 0a^2 + \cdots)$ and the polynomial expression $\sum_{i=0}^{n} f_i a^i$, we find that when $F[a] = F(a)$, the multiplicative inverse of any (nonzero) polynomial expression in a can be rewritten as a polynomial expression in a.

Continuing to study the subfield of K generated by F and a, two questions now spring to mind about the field $F(a)$. First, is there some intrinsic way of telling when $F[a] = F(a)$? That is, is there some property of the element a that determines whether $F[a]$ equals $F(a)$? Second, does the degree of $F(a)$ over F (that is, $[F(a) : F]$) depend in any way on the element a chosen? And if so, how? As it turns out, the same concept answers both questions, and this is the concept of the *minimal poylnomial* of a over F.

We will delay a full discussion of the minimal polynomial of an element until we have had a chance to review some facts about polynomials. However, as always, let us consider some examples that will be helpful. In fact, the two examples that we considered above, Example 4.7 and Example 4.8, will do quite nicely to illustrate the concept of the *polynomial* (if any) *satisfied by the element a over the field F*.

In Example 4.7, notice that the element $\sqrt{2}$ has the following property: there exist rational numbers q_0, q_1, and q_2 such that $q_2(\sqrt{2})^2 + q_1(\sqrt{2}) + q_0 = 0$. What are q_0, q_1, and q_2? This is easy: $q_0 = -2$, $q_1 = 0$, and $q_2 = 1$, that is, $\sqrt{2}$ satisfies $(\sqrt{2})^2 - 2 = 0$. We describe this by saying "$\sqrt{2}$ satisfies the polynomial (with rational coefficients) $t^2 - 2$." (The reason for this terminology is that when you substitute $\sqrt{2}$ for the variable t, you get zero. Note that we could have used any other variable to describe the polynomial. For instance, we could have said "$\sqrt{2}$ satisfies the polynomial $z^2 - 2$."

The only reason we did not use "x" as our variable above is to avoid confusion with the "x" of Example 4.4, where the basic objects of study are themselves all quotients of polynomials in the variable x.)

Now let us consider Example 4.8. Do there exist an integer n ($n \geq 0$) and real numbers $r_0, r_1, \ldots, r_n$, not all zero, such that the element x^2 satisfies the equation $r_n(x^2)^n + r_{n-1}(x^2)^{n-1} + \cdots + r_1(x^2) + r_0 = 0$? The answer is no! Remember, the "x" in $\mathbb{R}(x)$ is just a variable. Thus, $r_n(x^2)^n + r_{n-1}(x^2)^{n-1} + \cdots + r_1(x^2) + r_0$ is just a polynomial in x with real coefficients (except that all odd powers of x have coefficient 0). When will such a polynomial equal the zero polynomial? When all its coefficients are zero! Thus, the only way that x^2 can satisfy $r_n(x^2)^n + r_{n-1}(x^2)^{n-1} + \cdots + r_1(x^2) + r_0 = 0$ is if all the r_i are zero. We describe this by saying "x^2 does not satisfy any polynomial with coefficients in $\mathbb{R}$ except the zero polynomial." (Note that in the language of vector spaces, the set $\{1, x^2, x^4, x^6, \ldots\}$ is $\mathbb{R}$-linearly independent. Why?)

Motivated by these examples, we have the following:

Definition 4.10

Given a field extension K/F, an element a of K is said to be *algebraic over* F if there exist an integer n ($n \geq 0$) and elements $f_0, f_1, \ldots, f_n$ in F, *not all* f_i *being zero,* such that $f_na^n + f_{n-1}a^{n-1} + \cdots + f_1a + f_0 = 0$. (In such a situation, we say that the *element* a *satisfies the polynomial* $f_nt^n + f_{n-1}t^{n-1} + \cdots + f_1t + f_0$.) If no such integer n and no such elements $f_i \in F$ (again, not all f_i being zero) can be found, then a is said to be *transcendental over* F, and we say that a *does not satisfy any polynomial with coefficients in* F *except the zero polynomial.*

Note that if a is algebraic over F, then the polynomial with coefficients in F that a satisifies is not unique. For if a satisfies the polynomial $f(t) = f_nt^n + f_{n-1}t^{n-1} + \cdots + f_1t + f_0$ (that is, if $f_na^n + f_{n-1}a^{n-1} + \cdots + f_1a + f_0 = 0$) and if $g(t)$ is *any* nonzero polynomial with coefficients in F, then a satisfies the polynomial obtained by multiplying $f(t)$ and $g(t)$. (Why?) For instance, $\sqrt{2}$ satsifies $t^2 - 2$, but it also satisfies the polynomial $3(t^2 - 2)$ ($= 3t^2 - 6$), as well as the polynomial $(t^2 - 2)(t + 1)$ ($= t^3 + t^2 - 2t - 2$), and so on.

If K/F is a field extension, then every element of F is trivially algebraic over F. For, given $a \in F$, a satisfies the polynomial $x - a$, a nonzero polynomial *with coefficients in* F! Notice that we cannot extend this argument and claim that because every element a of K satisfies the polynomial $x - a$, every element of K must be algebraic over F. Why not? This is because when a is an element of K that is not in F, the polynomial $x - a$ does *not* have its coefficients in F!

Definition 4.11
In the special case where $F = \mathbb{Q}$ and $K = \mathbb{C}$, an element a in $\mathbb{C}$ that is algebraic over $\mathbb{Q}$ is simply referred to as an *algebraic number*. Similarly, an element a in $\mathbb{C}$ that is transcendental over $\mathbb{Q}$ is simply referred to as a *transcendental number*.

What are some examples of algebraic and transcendental numbers? We have already seen one algebraic number—$\sqrt{2}$. It is easy to write down several numbers that are obviously algebraic numbers: $\sqrt[3]{2}$, $\sqrt[4]{2}$, $\sqrt[19]{5}$, etc. Also, every rational number is an algebraic number. (Why?) Here are some less obvious algebraic numbers: $1 + \sqrt{2}$, $\sqrt{2} + \sqrt{3}$, $2 + \sqrt{3}\sqrt[3]{5}$, etc. (See the exercises.)

How about examples of transcendental numbers? It is quite difficult to pick a complex number and show that it is transcendental. Compare the process of showing that a certain number is algebraic with the process of showing that a certain number is transcendental. To show that a complex number a is algebraic, it is sufficient to exhibit *one* polynomial with rational coefficients that a satisfies. But to show that a given complex number a is transcendental, one needs to show that a does not satisfy *any* nonzero polynomial with rational coefficients. Since there are infinitely many polynomials with rational coefficients, showing that a given complex number is transcendental is obviously going to be quite difficult! It is not surprising, therefore, that the problem of showing that certain specific complex numbers are transcendental has occupied a number of prominent mathematicians for a long time. It has been known for quite a while, for instance, that both e and π are transcendental. (The fact that π is transcendental, for instance, will be central to one of the questions on constructibility that we will answer in Chapter 7.)

Strangely enough, while it is quite hard to *start* with a given complex number and prove that it is transcendental, it is somewhat easier to *exhibit* (from scratch) complex numbers that are transcendental. This is because of the existence of certain theorems in the subject that allow one to construct such numbers. One is a theorem of Liouville that, for instance, yields the transcendental number $\sum_{i=1}^{\infty} 2^{-i!}$ $(= 2^{-1} + 2^{-2} + 2^{-6} + 2^{-24} + \cdots)$. Another is the celebrated Gelfond-Schneider theorem, which (in a simplified form) states that if α is a real number, $\alpha \neq 0$, $\alpha \neq 1$, such that α is algebraic, and if β is a real number such that β is also algebraic but $\beta \notin \mathbb{Q}$, then α^β is transcendental. This allows us to exhibit several transcendental numbers, such as $2^{\sqrt{2}}$, $\sqrt{3}^{\sqrt{5}}$, etc.

If a is a transcendental number, must a^2 also be a transcendental number? How about $1/a$? (See Exercise 5. Also, see the notes on pages 116–117 for more remarks on algebraic and transcendental numbers.)

Now suppose K is an extension field of a field F of *finite* degree. The following theorem tells us that *every* element of K is algebraic over F. The proof is really quite cute! Notice that concepts from the theory of vector spaces are very intrinsic to the proof.

Theorem 4.12
Suppose K/F is a field extension with $[K : F]$ finite. Then every element of K is algebraic over F.

Proof Let $n = [K : F]$ (so n is finite). Let a be an arbitrary element of K. To prove that a is algebraic over F, we need to display an integer m $(m \geq 0)$ and elements $f_i \in F$, $i = 1, \ldots, m$, with not all $f_i = 0$, such that $f_m a^m + \cdots + f_1 a + f_0 = 0$. Consider the set $S = \{1, a, a^2, \ldots, a^n\}$. We may assume that the elements a^i $(i = 0, \ldots, n)$ are all distinct, since if, say $a^i = a^j$ for some i and j with $0 \leq i < j \leq n$, then a satisfies the polynomial $x^i - x^j$, and is therefore already algebraic over F. Now think of S as a set of vectors (in the F-vector space K). Suppose for the moment that these vectors in S are linearly independent over F. By Theorem 3.19, the number of elements in S must be at most n. On the other hand, we have seen above that we may assume that the elements a^i $(i = 0, \ldots, n)$ are all distinct, so S contains $n + 1$ elements—a contradiction. Hence, the elements of S must be linearly dependent over F.

But what does it mean for the elements of S to be linearly dependent over F? It means that there are scalars $f_0, f_1, \ldots, f_n$, not all zero, such that $f_0 \cdot 1 + f_1 \cdot a + \cdots + f_{n-1} \cdot a^{n-1} + f_n a^n = 0$. View this as an equation, not concerning vectors and scalars, but as one concerning elements of the field K: this says that there are elements $f_0, f_1, \ldots, f_n$ in F, not all zero, such that $f_n a^n + f_{n-1}a^{n-1} + \cdots + f_1 a + f_0 = 0$. This is precisely what it means for a to be algebraic over F! Since a was an arbitrary element of K, we have proved the theorem. □

Here is an example that illustrates the proof of this theorem.

Example 4.13

Take $F = \mathbb{Q}$ and $K = \mathbb{Q}(\sqrt{2})$. We know that $[K : F] = 2$. Suppose we wished to prove (without recourse to the statement of this theorem) that $1 + \sqrt{2}$ is algebraic over $\mathbb{Q}$. Consider the set $S = \{1, 1 + \sqrt{2}, (1 + \sqrt{2})^2\}$. Following the ideas in the proof of the theorem, S must be $\mathbb{Q}$-linearly dependent, since S has three elements while $[\mathbb{Q}(\sqrt{2}) : \mathbb{Q}] = 2$. Thus, there must exist rational numbers q_0, q_1, and q_2, not all zero, such that $q_2(1 + \sqrt{2})^2 + q_1(1 + \sqrt{2}) + q_0 = 0$. Can we guess what these three rational numbers must be? Notice that $(1 + \sqrt{2})^2 = 3 + 2\sqrt{2} = 2(1 + \sqrt{2}) + 1$. Hence, we find that $(1 + \sqrt{2})^2 - 2(1 + \sqrt{2}) - 1 = 0$, that is, $q_0 = -1$, $q_1 = -2$ and $q_2 = 1$. Described alternatively, $1 + \sqrt{2}$ satisfies the polynomial $x^2 - 2x - 1$.

Theorem 4.12 allows us to introduce one more concept. So far, given a field extension K/F, we have seen what it means for a *single element* $a \in K$ to be algebraic over F. Theorem 4.12 shows that if K is a finite-dimensional extension of F, then *every* element of K is algebraic over F. Motivated by this, let us make the following definition:

Definition 4.14

Let K/F be a field extension. Then K is said to be *algebraic over F* if *every* element of K is algebraic over F.

(Thus, what is new in this definition is that the notion of being algebraic is being applied to a whole field, not just to a single element.)

In particular, if K is a finite-dimensional extension of $\mathbb{Q}$, then Theorem 4.12 shows that K is algebraic over $\mathbb{Q}$. Finite-dimensional extensions of $\mathbb{Q}$ are called *algebraic number fields* and have been studied extensively.

Note that the converse of Theorem 4.12 is not true: it is possible for K to be algebraic over F and yet be infinite-dimensional over F! (See the remarks on algebraic numbers in the notes on page 116.)

Our next chapter will be a review of polynomials.

Exercises

1. Prove that if K/F is a field extension with $[K : F] = 1$, then K must equal F. (Hint: Assume that $K \neq F$. Then show that for any a in K that is not in F, a and 1 must be F-linearly dependent. Why is this a contradiction?) Now show that the converse is also true, that is, show that if $K = F$, then $[K : F] = 1$.

2. Suppose K/F is a field extension such that $[K : F] = p$ for some prime p. If L is a field such that $F \subseteq L \subseteq K$, prove that either $L = F$ or $L = K$. (Hint: What can you say about $[L : F]$ and $[K : L]$?)

3. Discover a polynomial with rational coefficients that the following complex numbers satisfy: (a) $1 + \sqrt{2}$, (b) $\sqrt{2} + \sqrt{3}$, (c) $\sqrt{2} + \sqrt{3} + \sqrt{6}$, (d) $2 + \sqrt{3}\sqrt[3]{5}$, and (e) $i + \sqrt{3}$.

4. Prove that if a is an algebraic number, then $1/a$ is also algebraic. (Hint: Consider any nonzero polynomial with rational coefficients that a satisfies. How can you manipulate this polynomial to arrive at a polynomial satisfied by $1/a$?)

5. Prove that if a is a transcendental number, then a^n is also transcendental for all nonzero integers n.

6. If a is a transcendental number, prove that $\mathbb{Q}[a]$ is not a field, so in particular, $\mathbb{Q}[a] \neq \mathbb{Q}(a)$. (Hint: Assume to the contrary that $\mathbb{Q}[a]$ is a field. Then the (nonzero) element a should have its inverse in $\mathbb{Q}[a]$. Hence, $1/a = c_0 + c_1a + \cdots + c_na^n$ for some integer n ($n \geq 0$) and some rational numbers $c_0, c_1, \ldots, c_n$. Multiply through by a and bring all terms to one side. Why do you arrive at a contradiction? Where in your proof are you using the assumption that a is transcendental over F?)

7. This exercise is designed to show that if a complex number α is algebraic over $\mathbb{Q}[\sqrt{2}]$, then it is also algebraic over $\mathbb{Q}$.

   (a) Given any element $x = a + b\sqrt{2}$ in $\mathbb{Q}[\sqrt{2}]$, let $\bar{x}$ denote the element $a - b\sqrt{2}$.

      i. Prove that for any $x \in \mathbb{Q}[\sqrt{2}]$ both $x + \bar{x}$ and $x\bar{x}$ are *rational* numbers.

      ii. Show that for any two elements x and y in $\mathbb{Q}[\sqrt{2}]$, $x\bar{y} + \bar{x}y$ is also a *rational* number.

   (b) Let $f(x) = f_n x^n + f_{n-1}x^{n-1} + \cdots + f_0$ be a polynomial with coefficients in $\mathbb{Q}[\sqrt{2}]$. Let $\overline{f_i}$ be as in part 7a above, and let $\overline{f(x)}$ denote the polynomial $\overline{f_n}x^n + \overline{f_{n-1}}x^{n-1} + \cdots + \overline{f_0}$, which is also a polynomial with coefficients in $\mathbb{Q}[\sqrt{2}]$. Prove that all the coefficients of the polynomial $f(x)\overline{f(x)}$ (that is, the polynomial obtained by multiplying out $f(x)$ and $\overline{f(x)}$) are *rational* numbers. (Hint: Look at the coefficients of the product $f(x)\overline{f(x)}$ and apply the results of part 7a. For instance, the highest coefficent is $f_n\overline{f_n}$, which by part 7(a)i, is a rational number.)

   (c) Now use the first part to prove the assertion at the beginning of this problem, that is, show that if a complex number α is algebraic over $\mathbb{Q}[\sqrt{2}]$, then it is also algebraic over $\mathbb{Q}$. (Hint: Apply the result of part 7b above to any polynomial with coefficients in $\mathbb{Q}[\sqrt{2}]$ satisfied by α.)

   A more general statement is true: If K is any subfield of $\mathbb{C}$ that is algebraic over $\mathbb{Q}$, and if a complex number α is algebraic over K, then α is also algebraic over $\mathbb{Q}$! You will be asked to prove this in the special case where $K/\mathbb{Q}$ is finite-dimensional in Exercise 6 in Chapter 6.

8. The following result finds use in algebraic geometry (where it is an ingredient in the proof of a theorem known as the "Going Up Theorem"). Let K be a field, and let R be a sub*ring* of K. Suppose that every element of K satisfies a polynomial whose coefficients are in R and *whose highest degree coefficient is* 1. Show that R is actually a sub*field* of K. (Hint: R is an integral domain—why?—so you only need to prove that for every nonzero $a \in R$, $1/a$ is also in R. The hypothesis shows that $1/a$ will satisfy a suitable polynomial with coefficients in R; try to manipulate this

polynomial. Where will your argument break down if the highest degree coefficient is not 1? Can you think of an example of a field K and subring R such that every element of K satisfies a polynomial with coefficients in R, yet R is not a field?)

9. We saw (without proof!) in the text that e is transcendental. It follows from this that e cannot be rational (why?). This exercise is designed to give a *direct* and elementary proof of the irrationality of e. Recall that e is defined by the infinite series $1 + \frac{1}{1!} + \frac{1}{2!} + \frac{1}{3!} + \cdots$.

   (a) Prove that for any integer $n \geq 1$, the infinite series $\frac{1}{(n+1)} + \frac{1}{(n+1)(n+2)} + \frac{1}{(n+1)(n+2)(n+3)} + \cdots$ is

      i. convergent, and

      ii. converges to a real number *strictly between* 0 and 1.

      (Hint: Notice that the given series is bounded term by term by the series $\frac{1}{(n+1)} + \frac{1}{(n+1)^2} + \frac{1}{(n+1)^3} + \cdots$.)

   (b) Assume to the contrary that e is rational, and write $e = m/n$, where m and n are integers and $n \neq 0$. Since the series for e shows that e is positive, we may assume that both m and n are positive, so in particular, we may assume $n \geq 1$. Prove that the series considered in part 9a above must converge to an integer. (Hint: Breaking up the infinite series for e into two parts, we find that $m/n = \sum_{i=0}^{n} 1/i! + \sum_{i=n+1}^{\infty} 1/i!$. Now multiply through by $n!$. What do you notice about $n! \cdot m/n$ and $n! \cdot \sum_{i=0}^{n} 1/i!$?)

   (c) Conclude from parts 9a and 9b above that e must be irrational.

10. This exercise deals with a very distinguished set of algebraic numbers, namely the various *roots of unity*. For any integer n, $n \geq 2$, let ω_n denote the complex number $\cos(2\pi/n) + i\sin(2\pi/n)$. (It would be helpful for you to locate ω_n on the complex plane for a few small values of n.)

    (a) For what values of n is ω_n a real number?

    (b) Both ω_2 and ω_4 are familiar complex numbers—what are they?

(c) Show that $\omega_n^n = 1$, but $\omega_n^k \neq 1$ for any integer k with $1 \leq k < n$. (Hint: Recall what De Moivre's theorem tells you about how complex numbers multiply—$\left(\cos(\theta) + i\sin(\theta)\right)^j = \cos(j\theta) + i\sin(j\theta)$.) The complex number ω_n is known as a *primitive nth root of unity*. (The word "primitive" refers to the fact that no positive power of ω_n smaller than n equals 1. "Unity," of course, is just an old word for 1!)

(d) Show that ω_n satisfies the polynomial $x^{n-1} + x^{n-2} + \cdots + x + 1$. (Hint: You know that ω_n satisfies $x^n - 1$. What do you remember about how $x^n - 1$ factors? If you do not remember any factorization of $x^n - 1$, use long division to divide $x^n - 1$ by $x - 1$. Now use the fact that $\omega_n \neq 1$.) ω_n is thus an algebraic number!

The various nth roots of unity are essential for determining the nth roots of complex numbers—see Exercise 6 in Chapter 5 ahead.

Notes

Remarks on Theorem 4.2 Most beginning students are very daunted by this theorem and its proof, they seem so...well, *confusing!* This is a very natural reaction, since you have not yet gotten used to switching back and forth between viewing field elements as just field elements, then viewing them as vectors, and then again viewing them as scalars! However, as with all abstract material, if you work constantly at trying to understand this new way of looking at things, you will eventually grasp it, and in fact, it will even seem very natural to you!

But this does not mean that you drop everything else and focus on understanding this theorem; you would be better off using a layered approach. At the first reading, concentrate solely on understanding what the statement of the theorem is saying; ignore the proof. Once you understand the statement, *you should move forward in the chapter,* simply using the theorem when it is necessary (for instance, in the exercises). Come back later to the proof, and this time around, concentrate perhaps on understanding why $[L : K]$ and $[K : F]$ are finite if $[L : F]$ is finite. Come back yet again, and maybe try to understand why $\mathcal{A}$ must have precisely mn elements. Your next attack should perhaps be on why the elements of $\mathcal{A}$ span L as an F-vector space, and finally, you should try to understand why the elements of $\mathcal{A}$ are F-linearly independent. At no point do you sacrifice the rest of the chapter—you are constantly moving forward in the chapter even as you are trying to understand this theorem. Eventually, however, the theorem and its proof (and, as well, the rest of the chapter!) will become clear to you.

The trick, in brief, is to feed yourselves difficult material only in small doses!

Remarks on Algebraic Numbers The theory of algebraic numbers is a vast area of mathematics that has been explored very deeply in the past two centuries. It is a topic that arises very naturally—solutions to polynomial equations (whose coefficients are rational numbers) are all algebraic numbers!

It is not too hard to prove (see Exercise 7 in Chapter 6) that if α and β are two algebraic numbers (with $\beta \neq 0$), then so are $\alpha \pm \beta$, $\alpha\beta$, and α/β. It follows from this that the set of algebraic numbers forms a field! Write $\overline{\mathbb{Q}}$ for this field. $\overline{\mathbb{Q}}$ is, of course, a subfield of $\mathbb{C}$ that contains $\mathbb{Q}$, and it has the following remarkable property: every polynomial equation with coefficients in $\overline{\mathbb{Q}}$ has a solution in $\overline{\mathbb{Q}}$! (You may know that $\mathbb{C}$ also has this property, but here is a smaller field than $\mathbb{C}$ that has such a property.)

Of course, although $\overline{\mathbb{Q}}$ is definitely smaller than $\mathbb{C}$ (it does not contain any transcendental numbers, for instance), it is still a *huge* field—it is infinite-dimensional over $\mathbb{Q}$. (Thus, $\overline{\mathbb{Q}}$ is a counterexample to the converse of Theorem 4.12, since every element of $\overline{\mathbb{Q}}$ is algebraic over $\mathbb{Q}$.) But then again, when looked at in a different light, $\overline{\mathbb{Q}}$ does not seem all that big—for those of you who are familiar with the notion of cardinality of sets, $\overline{\mathbb{Q}}$ has the same cardinality as $\mathbb{Q}$.

Let K be any finite extension of $\mathbb{Q}$ (in other words, using the terminology introduced in the text, let K be any *algebraic number field*). Associated to K in a canonical way is a certain subring of K, R_K, which contains $\mathbb{Z}$, and is known as *the ring of algebraic integers in K*. This ring has been studied quite extensively. A fundamental question about algebraic number fields is whether the notion of unique prime factorization (which is so central to $\mathbb{Z}$) continues to hold in the ring of algebraic integers R_K of a given algebraic number field K. There are some very familiar algebraic number fields K for which unique prime factorization does *not* hold in R_K! You have already seen an example of this in Exercise 10 of Chapter 2—the ring

$\mathbb{Z}[\sqrt{-5}]$ of that exercise is precisely the ring of algebraic integers of the field $\mathbb{Q}(\sqrt{-5})$. It was an old problem of Gauss to determine all the fields of the form $\mathbb{Q}(\sqrt{d})$ (for *negative* integers d) whose ring of algebraic integers admitted unique prime factorization. It is known now that the only such fields are those with $d = -1, -2, -3, -7, -11, -19, -43, -67, -163$.

Isn't this fascinating? If you agree, then you should pursue number theory further!

Remarks on Transcendental Numbers This may seem surprising: there are *tons* of transcendental numbers! To make this more precise, we need some notions from set theory, specifically those connected with the cardinality of sets. For those familiar with such notions, recall that the rationals form a *countable* set, while the complexes are *uncountable*. (A set S is said to be countable if there exists a bijection between the integers and S. Loosely speaking, this means that there are "as many" elements in S as there are integers. An uncountable set, on the other hand, is to be thought of intuitively as being "very much bigger" than the integers. Thus, there are "only as many" rational numbers as there are integers, but "lots and lots" of complex numbers!) Now recall what we said earlier. $\overline{\mathbb{Q}}$, the set of all algebraic numbers, has the same cardinality as $\mathbb{Q}$. This means that $\overline{\mathbb{Q}}$ is countable, that is, there are "only as many" algebraic numbers as there are integers. The entire set of complex numbers, on the other hand, is uncountable, that is, it is "very much bigger" than the integers. Since every complex number that is not algebraic is transcendental, cardinality arguments then show that the set of transcendental numbers is also uncountable, that is, the set of transcendental numbers is also "very much bigger" than the integers. In particular, the set of transcendental numbers is "very much bigger" than the set of algebraic numbers.

5 CHAPTER Polynomials

Let F be an arbitrary field. We will study the polynomial ring $F[x]$ in this chapter. We will consider what it means for one polynomial to divide another, and we will formalize the process of long division. We will consider the notion of irreducibility of polynomials, and we will see that $F[x]$ behaves remarkably like the integers when it comes to factorizations of polynomials, with irreducible polynomials playing the role of prime numbers. In fact, we will prove that every polynomial factors into a product of irreducible polynomials, and that the irreducible polynomials that occur in this factorization are unique in a suitable sense! We will end with a discussion on the number of roots of a given polynomial.

We remark that the degree of a nonzero polynomial in $F[x]$, as well as its highest coefficient, are defined exactly as in $\mathbb{R}[x]$ (Example 2.7.7). Also, just as in $\mathbb{R}[x]$, the degree of the zero polynomial is not defined. Finally, as described in the remarks on Example 2.7.7 on page 60 in the notes to Chapter 2, the degree of the product of two polynomials is just the sum of the degrees of each polynomial. In particular, $F[x]$ is an integral domain: if $f(x)$ and $g(x)$ are two nonzero polynomials, then $f(x)g(x)$ is a polynomial of degree equal to $deg(f(x)) + deg(g(x))$, so $f(x)g(x)$ cannot be the zero polynomial. There is a nice consequence of this: by Lemma 2.19, we can always cancel polynomials from both sides of an equation!

Let us start with the process of division. Let us invoke our experience with the set of polynomials with coefficients in the rationals, that is, $\mathbb{Q}[x]$. We are all familiar with the long division process. What do we do when we divide a polynomial $f(x) \in \mathbb{Q}[x]$ by another polynomial $g(x) \in \mathbb{Q}[x]$ using long division? We find a *quotient* $q(x)$ and a *remainder* $r(x)$, that is, we find polynomials $q(x)$ and $r(x)$ such that $f(x) = q(x)g(x) + r(x)$. For instance, when we divide $x^3 + x^2 + 1$ by $2x^2 - 1$, the quotient is $(1/2)x + 1/2$ and the remainder is $(1/2)x + 3/2$. (Check!) On the other hand, when we divide $x^2 - 1$ by $x^3 + x^2 + 1$, we observe that the degree of $x^2 - 1$ is less than the degree of $x^3 + x^2 + 1$, and we simply say that the quotient is 0 and the remainder is $x^2 - 1$ (that is, $x^2 - 1 = 0 \cdot (x^3 + x^2 + 1) + x^2 - 1$). As yet another example, when we divide $x^3 - 1$ by $x - 1$, we find that $x - 1$ divides *exactly* into $x^3 - 1$, with quotient $x^2 + x + 1$. We say that the remainder $r(x)$ in this situation is 0.

Why does this algorithm work? In other words, given polynomials $f(x)$ and $g(x)$, why do the two polynomials yielded by the algorithm—the one at the top of the long division table, "$q(x)$," and the one at the bottom of the long division table, "$r(x)$"—satisfy the relation $f(x) = q(x)g(x) + r(x)$? Also, what can we say about $r(x)$ in relation to $g(x)$?

To answer these questions, let us take a specific example: let us consider the division of $x^4 + x^3 + x^2 + x + 1$ by $2x^2 + 1$.

Example 5.1

$$
\begin{array}{rrrrrrrrrr}
 & & \frac{1}{2}x^2 & + & \frac{1}{2}x & + & \frac{1}{4} & & & \\
\cline{3-10}
2x^2 + 1 & \big) & x^4 & + & x^3 & + & x^2 & + & x & + \; 1 \\
 & & x^4 & + & & & \frac{1}{2}x^2 & & & \\
\cline{5-10}
 & & & & x^3 & + & \frac{1}{2}x^2 & + & x & + \; 1 \\
 & & & & x^3 & & & + & \frac{1}{2}x & \\
\cline{7-10}
 & & & & & & \frac{1}{2}x^2 & - & \frac{1}{2}x & + \; 1 \\
 & & & & & & \frac{1}{2}x^2 & & & + \; \frac{1}{4} \\
\cline{9-10}
 & & & & & & & & \frac{-1}{2}x & + \; \frac{3}{4}
\end{array}
$$

When did we stop? We stopped when the result of the subtraction (that we do at every stage) had a degree *less* than that of $2x^2 + 1$, and we called the result of this last subtraction (namely $(-1/2)x + (3/4)$) the remainder.

Also, do the two polynomials yielded by this long division—$(1/2)x^2 + (1/2)x + (1/4)$ and $(-1/2)x + (3/4)$—satisfy the relation

$$x^4 + x^3 + x^2 + x + 1 = \big((1/2)x^2 + (1/2)x + (1/4)\big)(2x^2 + 1) + \big((-1/2)x + (3/4)\big)?$$

This can be verified by direct multiplication, but here is another way of seeing it. Looking at the polynomials involved in the first stage of the long division process, we have

$$\begin{aligned}(\frac{1}{2}x^2)(2x^2 + 1) &= x^4 + \frac{1}{2}x^2 \\ &= (x^4 + x^3 + x^2 + x + 1) - (x^3 + \frac{1}{2}x^2 + x + 1),\end{aligned}$$

where the last equality follows from the fact that $(x^4 + x^3 + x^2 + x + 1) - \big(x^4 + (1/2)x^2\big) = (x^3 + (1/2)x^2 + x + 1)$. We rearrange this to read

$$x^4 + x^3 + x^2 + x + 1 = \left(\frac{1}{2}x^2\right)(2x^2 + 1) + (x^3 + \frac{1}{2}x^2 + x + 1). \quad (5.1)$$

Now work inductively. By the same sort of an argument, we find from the second stage that

$$x^3 + \frac{1}{2}x^2 + x + 1 = \left(\frac{1}{2}x\right)(2x^2 + 1) + \left(\frac{1}{2}x^2 - \frac{1}{2}x + 1\right). \quad (5.2)$$

Similarly, we find from the third stage that

$$\frac{1}{2}x^2 - \frac{1}{2}x + 1 = \frac{1}{4}(2x^2 + 1) + \left(\frac{-1}{2}x + \frac{3}{4}\right). \quad (5.3)$$

Now add the three equations 5.1, 5.2, and 5.3 together. What do we find? The last term of equation 5.1 cancels with the left-hand side of equation 5.2, and the last term of equation 5.2 cancels with the left-hand side of equation 5.3! Collecting all the terms multiplying $2x^2 + 1$ into one polynomial, we find that $x^4 + x^3 + x^2 + x + 1 = \big((1/2)x^2 + (1/2)x + (1/4)\big)(2x^2 + 1) + \big((-1/2)x + (3/4)\big)$!

In general, when we divide a polynomial $f(x)$ by $g(x)$ by using long division, the polynomial that we get at every stage as a result of the subtraction step will have degree *less* than the corresponding polynomial at the previous stage. Since the degree decreases at every stage, it is clear that after a certain number of stages, the polynomial that results from the subtraction step *must either have degree less than that of $g(x)$ or must become zero.* We stop the long division process at this stage and call the result of this last subtraction the remainder $r(x)$. We call the polynomial that we obtain at the top of the long division table the quotient $q(x)$. In particular, if $f(x)$ already has degree less than that of $g(x)$, we stop immediately, setting $q(x) = 0$ and $r(x) = f(x)$. The polynomials $q(x)$ and $r(x)$ will satisfy the relation $f(x) = q(x)g(x) + r(x)$. (This is of course clear if $deg(f(x)) < deg(g(x))$ since $q(x) = 0$ and $r(x) = f(x)$ in this case. For the general case, the relation $f(x) = q(x)g(x) + r(x)$ can be established by an argument exactly as in Equations 5.1, 5.2, and 5.3.)

We have thus seen that if $f(x)$ and $g(x)$ are two polynomials in $\mathbb{Q}[x]$ with $g(x) \neq 0$, then using long division, one can always find polynomials $q(x)$ and $r(x)$ in $\mathbb{Q}[x]$ such that $f(x) = q(x)g(x) + r(x)$ with either $r(x) = 0$, or $0 \leq deg(r(x)) < deg(g(x))$.

Now a natural question arises. Given $f(x)$ and $g(x)$ as above, the division algorithm certainly yields polynomials $q(x)$ and $r(x)$ such that $f(x) = q(x)g(x) + r(x)$ with either $r(x) = 0$, or $0 \leq deg(r(x)) < deg(g(x))$. But perhaps by an alternative process, could we not have found two *different* polynomials, say $q'(x)$ and $r'(x)$, such that $f(x) = q'(x)g(x) + r'(x)$ with either $r'(x) = 0$, or $0 \leq deg(r'(x) < deg(g(x))$? The answer is no! To see why this must be so, suppose that an alternative process did yield polynomials $q'(x)$ and $r'(x)$ with the stated properties. Then

$$f(x) = q(x)g(x) + r(x) = q'(x)g'(x) + r'(x),$$

so

$$(q'(x) - q(x))g(x) = r(x) - r'(x). \tag{5.4}$$

Now suppose $q'(x) \neq q(x)$. Then $q(x) - q'(x)$ is a nonzero polynomial, so $(q'(x) - q(x))g(x) \neq 0$, and moreover, the degree of $(q'(x) - q(x))g(x)$ is at least equal to the degree of $g(x)$. (Why?) On the other hand, each of $r(x)$ and $r'(x)$ has degree less than that of $g(x)$ or is the zero polynomial, so $r(x) - r'(x)$ also has degree less than that of $g(x)$ or is the zero polynomial. (Check!). Since $(q'(x) - q(x))g(x)$ has to equal $r(x) - r'(x)$, this is a contradiction. Hence $q'(x)$ must equal $q(x)$, and it then follows from Equation 5.4 that $r'(x)$ must also equal $r(x)$.

We describe the fact that $q'(x)$ must equal $q(x)$ and $r'(x)$ must equal $r(x)$ by saying that $q(x)$ and $r(x)$ are the *unique* polynomials that satisfy the relation $f(x) = q(x)g(x) + r(x)$ and have the property that either $r(x) = 0$ or $0 \leq deg(r(x) < deg(g(x))$.

So far, we have been working with polynomials whose coefficients are rational numbers. What if we had an arbitrary field F, and we considered polynomials $f(x)$ and $g(x)$ with coefficients in F? Would we still be able to find a quotient polynomial $q(x)$ and a remainder $r(x)$? The answer is yes—one would just have to perform the same sort of long division that we do with polynomials in $\mathbb{Q}[x]$, except that all the coefficients would now come from the field F. As with $\mathbb{Q}[x]$, we would stop the long division process when the result of the subtraction step either has degree less than that of $g(x)$ or is zero, and we would call the result of this last subtraction the remainder $r(x)$. We would call the polynomial that we would obtain at the top of the table the quotient $q(x)$. The same arguments as in the case of $\mathbb{Q}[x]$ would show that the polynomials $q(x)$ and $r(x)$ satisfy $f(x) = q(x)g(x) + r(x)$ and are unique.

We summarize these discussions in the following:

Theorem 5.2 (Division Algorithm)
Let F be an arbitrary field and $f(x)$ and $g(x)$ be in $F[x]$, with $g(x) \neq 0$. Then there exist unique polynomials $q(x)$ and $r(x)$ in $F[x]$ with either $r(x) = 0$ or $0 \leq deg(r(x)) < deg(g(x))$, such that $f(x) = q(x)g(x) + r(x)$.

(See the notes on page 147 for a discussion on the division algorithm in $R[x]$, where R is an arbitrary ring.)

Now let us consider a special case of the division algorithm that will be quite useful. This is the case where the polynomial $g(x)$ is of the form $x - a$ for some a in F. But first, we will legitimize the process of substitution of values for x in the following lemma.

Lemma 5.3 (Evaluation Homomorphism)
Let $f(x)$ and $g(x)$ be arbitrary polynomials in $F[x]$, and let $p(x) =$

$f(x)g(x)$ and let $q(x) = f(x) + g(x)$. Then, for any a in F, we have $p(a) = f(a)g(a)$ and $q(a) = f(a) + g(a)$.

(Here, if $f(x) = \sum_{i=0}^{n} f_i x^i$, then $f(a)$ is the element obtained by substituting a for x in $f(x)$, that is, $f(a) = \sum_{i=0}^{n} f_i a^i$. Note that since each term $f_i a^i$ is in F, $f(a) = \sum_{i=0}^{n} f_i a^i$ *is an element of* F.)

The proof is easy and will be assigned to you as Exercise 1 at the end of this chapter. What this lemma says is that substituting a for x in the product (or sum) of two polynomials is the same as substituting a for x in each of the two polynomials and multiplying (or adding) the results together. Furthermore, this lemma can be used repeatedly (and we will do so below). Let $f(x)$, $g(x)$, and $h(x)$ be three polynomials, and let $p(x) = f(x)g(x) + h(x)$. Write $q(x)$ for $f(x)g(x)$, so $p(x) = q(x) + h(x)$. The lemma shows that $p(a) = q(a) + h(a)$. Applying the lemma once again to $q(x)$, $q(a) = f(a)g(a)$. Putting this together, we find that $p(a) = f(a)g(a) + h(a)$.

Now let us revert to the division algorithm and consider the result of dividing a polynomial $f(x)$ by a polynomial of the form $x - a$, where a is an arbitrary element of F. The algorithm tells us that there exist unique polynomials $q(x)$ and $r(x)$ with $r(x) = 0$ or with $0 \leq deg(r(x)) < deg(g(x))$ such that $f(x) = q(x)(x - a) + r(x)$. Notice that the remainder $r(x)$ must be a constant. (Why? Can the constant equal zero?) Substituting $x = a$, the discussion following Lemma 5.3 above shows us that $f(a) = q(a)(a - a) + r(a)$, that is, $f(a) = r(a)$. Since $r(x)$ is just a constant, $r(a)$ is just $r(x)$ (there is no "x" in $r(x)$ for us to substitute a for!). In other words, $f(a)$ is precisely the remainder $r(x)$. We summarize this in the following corollary to the division algorithm:

Corollary 5.4

Let $f(x)$ be a polynomial in $F[x]$, and let a be an element of F. Then the remainder obtained by dividing $f(x)$ by $x - a$ is just $f(a)$.

Definition 5.5

Given a nonzero polynomial $f(x) \in F[x]$, an element $a \in F$ is said to be a *root* of $f(x)$ if $f(a) = 0$.

Observe that Corollary 5.4 above shows that if a is a root of $f(x)$, then $f(x)$ must equal $(x-a)q(x)$ for a suitable polynomial $q(x)$. Conversely, if $f(x) = (x-a)q(x)$ for some polynomial $q(x) \in F[x]$, then $f(a) = (a-a)q(a) = 0$, so a is a root of $f(x)$.

Example 5.6

If F is any field, then every linear polynomial in $F[x]$ has *exactly* one root in F. For let the linear polynomial be given by $f(x) = ax + b$, where $a \neq 0$. Then $f(-b/a) = 0$, so $-b/a$ is a root of $f(x)$, and of course, $-b/a \in F$. Conversely, let $t \in F$ be a root of $f(x)$. Then $f(x) = ax + b = (x-t)g(x)$ for some $g(x) \in F[x]$. Comparing degrees, we find that $g(x)$ is a constant, say g_0. Comparing the highest coefficients on both sides, we find that $g_0 = a$. Next, comparing the constant terms on both sides, we find that $-tg_0 = b$, and so $t = -g_0/b = -a/b$. Thus, $-a/b$ is the only root of $ax + b$.

(Can a nonzero constant polynomial have a root?)

More generally, let us consider the case where $g(x)$ is an arbitrary polynomial and where the remainder obtained by dividing $f(x)$ by $g(x)$ is zero.

Definition 5.7

Given polynomials $f(x)$ and $g(x)$ in $F[x]$ with $g(x) \neq 0$, we say that $g(x)$ *divides* $f(x)$, or $g(x)$ is a *factor* of $f(x)$, if $f(x) = q(x)g(x)$ for some polynomial $q(x)$ in $F[x]$.

(For instance, if a is a root of $f(x)$, then our discussions above show that $x-a$ divides $f(x)$.)

Notice that the definition of divisibility we gave above for polynomials is identical to the corresponding definition of divisibility in the integers (Definition 1.1 of Chapter 1). This should not be surprising. After all, this definition captures our intuitive notion of what it means for one element to divide another, and this intuitive notion is the same, whether we are working with the integers or are working with polynomials over a field. More generally, the same definition of divisibility would apply to *any* commutative ring—if R is any commutative ring, we would say that a nonzero element d divides an element a if there exists another element $b \in R$ such that $db = a$. (Why is commutativity an issue? How would you modify the notion of divisibility if your ring were not commutative?) This illustrates the general principle alluded to at the beginning of Chapter 2 (see page 29), that abstraction codifies phenomena that occur simultaneously in several mathematical sets.

For instance, the following lemma should remind you of Lemma 1.2 in Chapter 1.

Lemma 5.8

If $d(x)$ is a nonzero polynomial such that $d(x)|a(x)$ and $d(x)|b(x)$ for two polynomials $a(x)$ and $b(x)$, then for any two polynomials $p(x)$ and $q(x)$, $d(x)|(p(x)a(x) + q(x)b(x))$.

The proof is identical to the proof of Lemma 1.2 in Chapter 1, and we will omit it. The lemma applies to any commutative ring—if R is any commutative ring and if a nonzero element d of R divides two elements a and b of R, then d divides $pa + qb$ for all p and q in R.

To develop a theory of factorization of polynomials, we need to develop first the concept of the greatest common divisor of two polynomials. Let us review how we developed the greatest common divisor in the case of the integers. Given two nonzero integers a and b, we observed (see the discussion just before Definition 1.5 in Chapter 1) that the number of common divisors of a and b is *finite*, and because of this finiteness, we were guaranteed that there would be one common divisor that would be larger than all the rest. We defined this particular (largest) common divisor to be the greatest common divisor of a and b.

In the case of polynomials, there is no good notion of one polynomial being larger than another polynomial (see the notes on page 148 for some remarks this), so we cannot define the greatest common divisor of two polynomials to be the largest of all the common divisors. What is the alternative?

As described in the remarks on the greatest common divisor on page 27 in the notes to Chapter 1, it is possible to turn Corollary 1.7 around and use it to define the greatest common divisor of two nonzero integers a and b as that positive integer d such that d is a common divisor of a and b, and such that every common divisor of a and b divides d. Of course, we ourselves did not go this route while defining the greatest common divisor of two integers, but we observed on page 27 in the notes to Chapter 1 that many textbooks use this definition, and that they do so because it generalizes (with a minor modification) to other number systems where a notion of "largest" common divisor may not exist. The ring $F[x]$ is certainly one place where a good notion of "largest" common divisor does not exist, and we will hence define the greatest common divisor as follows:

Definition 5.9

Given two nonzero polynomials $f(x)$ and $g(x)$ in $F[x]$, a polynomial $d(x)$ is said to be a *greatest common divisor (g.c.d.)* of $f(x)$ and $g(x)$ if

1. $d(x)$ divides both $f(x)$ and $g(x)$, and
2. if $h(x)$ is any polynomial that divides both $f(x)$ and $g(x)$, then $h(x)$ divides $d(x)$.

(See the notes on page 148 for some remarks on an alternative definition of the greatest common divisor.)

Notice that we left out the stipulation that the greatest common divisor be positive. Once again, this is because there is no good notion of one polynomial being greater than the zero polynomial (which is exactly it would mean for a polynomial to be positive).

Notice, too, that this definition does not address the issues of the existence and the uniqueness of the greatest common divisor of two polynomials. (How do we know that there exists a common divisor of $f(x)$ and $g(x)$ that is divisible by every other common divisor? Or for that matter, what if it turns out that there are several common divisors of $f(x)$ and $g(x)$, each divisible by every other common divisor?)

Let us consider the uniqueness question: If $d(x)$ and $d'(x)$ are two greatest common divisors of $f(x)$ and $g(x)$, is there a connection between $d(x)$ and $d'(x)$? Must $d(x)$ *equal* $d'(x)$? Well, by the first part of Definition 5.9, $d(x)$ and $d'(x)$ each divide both $f(x)$ and $g(x)$. Hence, by the second part of the same definition, $d(x)$ (being a common divisor of $f(x)$ and $g(x)$) must divide $d'(x)$, and $d'(x)$ (being a common divisor of $f(x)$ and $g(x)$) must divide $d(x)$. Hence $d'(x) = d(x)q(x)$ for some $q(x) \in F[x]$, and $d(x) = d'(x)q'(x)$ for some $q'(x) \in F[x]$. Thus, $d(x) = d(x)q(x)q'(x)$, or $1 = q(x)q'(x)$. This forces both $q(x)$ and $q'(x)$ to have degree 0 (why?), so $q(x)$ and $q'(x)$ must be constants! Thus, $d'(x)$ must just be a constant times $d(x)$, or in other words, any two greatest common divisors of $f(x)$ and $g(x)$ must only differ by a constant factor.

Conversely, it easy to see that if $d(x)$ is a greatest common divisor of $f(x)$ and $g(x)$, and if c is any nonzero element of F, then $cd(x)$ is also a greatest common divisor of $f(x)$ and $g(x)$. Let us check this. We need to verify that $cd(x)$ also satsifes Definition 5.9. Since $d(x)$ is a greatest common divisor of $f(x)$ and $g(x)$, $d(x)|f(x)$, and consequently, $f(x) = d(x)p(x)$ for some polynomial $p(x)$. We rewrite this as $f(x) = \left(cd(x)\right)\left(c^{-1}p(x)\right)$, which shows that $cd(x)|f(x)$. We can similarly see that $cd(x)|g(x)$, so $cd(x)$ is indeed a common divisor of $f(x)$ and $g(x)$. Now suppose that $h(x)$ is a common divisor of $f(x)$ and $g(x)$. Then $h(x)$ must divide $d(x)$ since $d(x)$ is a greatest common divisor of $f(x)$ and $g(x)$, and hence $d(x) = h(x)q(x)$ for some polynomial $q(x)$. We rewrite this as $cd(x) = h(x)\left(cq(x)\right)$, which shows that $h(x)$ divides $cd(x)$. Thus, $cd(x)$ satisfies Definition 5.9 and is also a greatest common divisor of $f(x)$ and $g(x)$.

What this discussion shows is that *the greatest common divisor of two nonzero polynomials is not truly unique, it is only determined up to a constant multiple.* We describe this by saying that the greatest common divisor of two nonzero polynomials is *unique up to constant multiples,* and we often talk of *the* greatest common divisor of two polynomials, with the understanding that we are only specifying it up to a constant multiple.

We still have the first question to settle—must a greatest common divisor always exist? The following theorem, which is quite similar to Theorem 1.6, shows that the answer is yes!

Theorem 5.10

Let $f(x)$ and $g(x)$ be two nonzero polynomials in $F[x]$. Let $S = \{a(x)f(x) + b(x)g(x) \mid a(x),\ b(x) \in F[x],\text{ and } a(x)f(x) + b(x)g(x) \neq 0\}$. Let $d(x)$ be any element of S of least *degree. Then $d(x)$ is a g.c.d. of $f(x)$ and $g(x)$. Moreover, every element of S is divisible by $d(x)$.*

Proof Observe first that S is nonempty—it contains both $f(x)$ and $g(x)$ (why?). Now let $d(x)$ be any element of S of least degree. We need to show that $d(x)$ satisfies the two requirements for being a greatest common divisor of $f(x)$ and $g(x)$.

First, since $d(x)$ is an element of S, there exist polynomials $p(x)$ and $q(x)$ in $F[x]$ such that $d(x) = p(x)f(x) + q(x)g(x)$. Suppose that $d(x)$

does not divide $f(x)$. Then, by the division algorithm, we can write $f(x) = h(x)d(x) + r(x)$ for some polynomials $h(x)$ and $r(x)$ in $F[x]$ with $0 \leq deg(r(x)) < deg(d(x))$ and $r(x) \neq 0$. (Why is $r(x) \neq 0$?) Writing $d(x)$ as $p(x)f(x) + q(x)g(x)$, we find that $r(x) = (1 - p(x)h(x))f(x) + (-q(x)h(x))g(x)$. Taking $a(x) = 1 - p(x)$ and $b(x) = -q(x)$, we find that $r(x)$ is of the form $a(x)f(x) + b(x)g(x)$, and since $r(x) \neq 0$, $r(x)$ satisfies the requirements for being a member of S. But $r(x)$ has a lower degree than $d(x)$, and $d(x)$ was a polynomial in S of least degree. Since this is a contradiction, $d(x)$ must divide $f(x)$. A similar argument shows that $d(x)$ must divide $g(x)$ as well.

Thus $d(x)$ satisfies the first requirement for being a greatest common divisor of $f(x)$ and $g(x)$. Now suppose $e(x)$ is any polynomial in $F[x]$ that divides both $f(x)$ and $g(x)$. Then $f(x) = e(x)s(x)$ and $g(x) = e(x)t(x)$ for suitable polynomials $s(x)$ and $t(x)$ in $F[x]$. Then $d(x) = p(x)f(x) + q(x)g(x) = e(x)\big(p(x)s(x) + q(x)t(x)\big)$, so $e(x)$ divides $d(x)$. Thus, $d(x)$ is indeed a greatest common divisor of $f(x)$ and $g(x)$.

For the last statement of the proof, observe that since $d(x)$ divides both $f(x)$ and $g(x)$ by the first part of the theorem, $f(x) = h(x)d(x)$ and $g(x) = k(x)d(x)$ for suitable polynomials $h(x)$ and $k(x)$ in $F[x]$. Hence, every $a(x)f(x) + b(x)g(x)$ in S can be written as $d(x)\big(a(x)h(x) + b(x)k(x)\big)$ and is hence divisible by $d(x)$. □

For example, $x - 1$ is a g.c.d. of $x^2 - 1$ and $x^3 - 1$. Let us see how this follows from the theorem above. Note that $x - 1 = (1) \cdot (x^3 - 1) + (-x) \cdot (x^2 - 1)$. Does this make $x - 1$ a g.c.d. of $x^2 - 1$ and $x^3 - 1$? Not yet. We need to show that $x - 1$ has the *least* degree in the set S of the theorem. Suppose $x - 1$ is not of least degree. Then since $deg(x - 1) = 1$, the least degree in S can only be zero, that is, "the" g.c.d. of $x^2 - 1$ and $x^3 - 1$ must be 1. But note that $x - 1$ divides both $x^2 - 1$ and $x^3 - 1$. This means that $x - 1$ must divide the g.c.d, so the g.c.d. cannot be a constant. Hence the least degree in S must indeed be 1, so $x - 1$ is indeed "the" g.c.d. of $x^2 - 1$ and $x^3 - 1$.

We remarked in the paragraph above that even though a polynomial $h(x)$ may be expressible as $a(x)f(x) + b(x)g(x)$ for suitable polynomials $a(x)$ and $b(x)$, we *cannot* conclude from this that $h(x)$ must be the greatest common divisor of $f(x)$ and $g(x)$, since for all we know, there may be a polynomial of lower degree in the set S of Theorem 5.10 above. Now here is a question: if you know that the constant polynomial 1 is in the set S, can you conclude that the greatest common divisor of $f(x)$ and $g(x)$ is 1?

This immediately leads to the analog of yet another concept from the integers.

Definition 5.11
Two polynomials $f(x)$ and $g(x)$ in $F[x]$ are said to be *relatively prime* if 1 is a g.c.d. of $f(x)$ and $g(x)$.

Note that 1 is a g.c.d. of $f(x)$ and $g(x)$ if and only if c is a g.c.d of $f(x)$ and $g(x)$, where c is any nonzero constant. (Why?)

We immediately have the following:

Corollary 5.12
Two polynomials $f(x)$ and $g(x)$ in $F[x]$ are relatively prime if and only if there exist polynomials $p(x)$ and $q(x)$ in $F[x]$ such that $p(x)f(x) + q(x)g(x) = 1$.

Proof By now, you should be able to prove this yourselves! (See the question two paragraphs before Definition 5.11 above.) □

For instance, $x-1$ and x are relatively prime, since $1 \cdot (x) + (-1) \cdot (x-1) = 1$. Similarly, x^2+1 and x^3-1 are relative prime since $(x^4 - x^2 + 1)(x^2 + 1) + (-x^3 - 1)(x^3 - 1) = 2$. (Why is it sufficient to get 2 on the right-hand side to conclude that $x^2 + 1$ and $x^3 - 1$ are relatively prime?)

We will find the following useful. Notice the similarity with Lemma 1.10 of Chapter 1.

Lemma 5.13
If $f(x)$, $g(x)$, and $h(x)$ are three polynomials in $F[x]$ such that $f(x)|g(x)h(x)$ and $gcd(f(x), g(x)) = 1$, then $f(x)|h(x)$.

Proof Study the proof of Lemma 1.10 and furnish a proof yourselves! □

The next step in our quest for a theory of factorization of polynomials is to develop the analog of prime numbers. We address this now.

Let us start with an example. Consider the polynomial $f(x) = 2x + 1$ in $\mathbb{Q}[x]$. Suppose we could factor it as $f(x) = g(x)h(x)$ for some polynomials $g(x)$ and $h(x)$ in $\mathbb{Q}[x]$. Since the degrees of $g(x)$ and $h(x)$

must add up to the degree of $f(x)$, and since the degree of $f(x)$ is 1, we find either $g(x)$ must have degree 0 or $h(x)$ must have degree 0. What are the polynomials of degree 0? They are the nonzero constants! Thus, the only way we can factor $2x + 1$ is by having one of the factors be a constant. (As examples of such factorizations, we have $2x + 1 = 2(x + 1/2)$, or, $2x + 1 = (1/2)(4x + 2$, etc.) Notice that if one of the factors is a constant, then the other factor must have the same degree as $f(x)$, and this is indeed borne out by the two examples above.

Now, the idea behind factorization is to take a polynomial and write it as a product of polynomials that are in some sense *simpler* than the original polynomial. What could we possibly mean by "simpler?" Certainly one way in which the factors could be simpler than the original polynomial is if they were of smaller degree than the original polynomial. In the case of the polynomial $2x + 1$ above, we found that no matter which factorization we choose, one of the factors must always have the same degree as $2x + 1$, since the other factor must always be a constant. Such a factorization is really not of much help, since effectively, all we have done is factored a constant out of $2x + 1$. We think of this as a *trivial* factorization, and we say that $2x + 1$ can only be factored *trivially*. (This is analogous to the following situation in the integers: 2 can be factored as either $1 \cdot 2$ or $-1 \cdot -2$, but neither of these factorizations breaks 2 down into a product of "simpler" integers. In other words, these are just "trivial" factorizations of 2.) There is a concept behind this phenomenon:

Definition 5.14

We say that a *nonconstant* polynomial $f(x)$ in $F[x]$ is *irreducible* if whenever $f(x) = g(x)h(x)$ for some $g(x)$ and $h(x)$ in $F[x]$, then either $g(x)$ or $h(x)$ must be a constant. (A factorization into two polynomials, one of which is just a constant, is known as a *trivial factorization.*) Similarly, we say that $f(x)$ is *reducible* if $f(x) = g(x)h(x)$ for some $g(x)$ and $h(x)$ in $F[x]$, where neither $g(x)$ nor $h(x)$ is a constant. (A factorization into two polynomials, neither of which is a constant, is known as a *nontrivial factorization.*)

Our example above shows that the polynomial $2x + 1$ in $\mathbb{Q}[x]$ is irreducible. More generally, any polynomial of degree 1 is irreducible. (Why?) On the other hand, the polynomial $x^2 - 1$ in $\mathbb{Q}[x]$ is reducible, since $x^2 - 1$ factors as $(x-1)(x+1)$, a nontrivial factorization. Irreducible polynomials will turn out to be the analog of prime numbers. We will soon see that every polynomial can be factored into a product of irreducibles.

Notice that the definition of irreducibility only applies to nonconstant polynomials. We do not think of the constant polynomials as either reducible or irreducible. (This is analogous to the following situation with the integers—we think of 1 neither as a prime nor as a composite.)

An important feature of irreducibility comes to light when we have a field extension K/F and we consider polynomials in $F[x]$. Here is a specific example: Consider the field extension $\mathbb{Q}(\sqrt{2})/\mathbb{Q}$, and consider the polynomial $x^2 - 2$. Since $x^2 - 2$ has its coefficients in $\mathbb{Q}$, it clearly is an element of $\mathbb{Q}[x]$. On the other hand, since $\mathbb{Q} \subseteq \mathbb{Q}(\sqrt{2})$, we can consider the coefficients of $x^2 - 2$ as elements of $\mathbb{Q}(\sqrt{2})$, so $x^2-2 \in \mathbb{Q}(\sqrt{2})[x]$ as well. Now, as an element of $\mathbb{Q}[x]$, x^2-2 is irreducible (see Example 5.15.3 below). However, when viewed as an element of $\mathbb{Q}(\sqrt{2})[x]$, $x^2 - 2$ becomes reducible—it factors as $(x - \sqrt{2})(x+\sqrt{2})$! Thus, *reducibility and irreducibility are only defined within the context of a given field F*, and as this example shows, if K/F is a field extension, then the same polynomial $f(x)$ with coefficients in F may be irreducible as an element of $F[x]$, but may become reducible as an element of $K[x]$. To emphasize the field that we are considering, we often say in the situation above that "$f(x)$ is irreducible in $F[x]$," or "$f(x)$ is irreducible over F." Similarly, we say "$f(x)$ is reducible in $F[x]$," or "$f(x)$ is reducible over F."

Let us consider some examples of reducible and irreducible polynomials:

Examples 5.15

1. If F is *any* field, then every polynomial of degree 1 with coefficients in F is irreducible over F. (The same argument we used above for the irreducibility of $2x + 1$ over $\mathbb{Q}$ works for any field.)
2. Let F be any field and $f(x) \in F[x]$ be any polynomial of degree *greater than* 1. If $f(x)$ has a root in F, then $f(x)$ is reducible. For if $a \in F$ is a root of $f(x)$, then by Corollary 5.4, $f(x) = (x-a)g(x)$ for some $g(x) \in F[x]$. Since $deg(f(x)) > 1$ and $deg((x-a)) +$

$deg(g(x)) = deg(f(x))$, $deg(g(x)) \geq 1$, so we have a nontrivial factorization of $f(x)$.

3. The polynomial $x^2 - 2$ is irreducible in $\mathbb{Q}[x]$. For suppose it is reducible. Then $x^2 - 2 = g(x)h(x)$ for some polynomials $g(x)$ and $h(x)$ in $\mathbb{Q}[x]$, with neither $g(x)$ nor $h(x)$ being constants. Both $g(x)$ and $h(x)$ must hence be of degree 1. (Why?) Write $g(x) = ax + b$ and $h(x) = cx + d$ for suitable a, b, c, and d in $\mathbb{Q}$. (Why must a and c both be nonzero?) Since $\sqrt{2}$ satisfies $x^2 - 2 = 0$, we find, on substituting $\sqrt{2}$ for x, that $(a\sqrt{2} + b)(c\sqrt{2} + d) = 0$. Thus, either $a\sqrt{2} + b = 0$ or $c\sqrt{2} + d = 0$. Why does this lead to a contradiction?
4. In Example 3 above, the argument that $x^2 - 2$ is irreducible over $\mathbb{Q}$ ultimately invoked the fact that $\sqrt{2} \notin \mathbb{Q}$. In other words, the irreducibility of $x^2 - 2$ over $\mathbb{Q}$ ultimately followed from the fact that the polynomial $x^2 - 2$ has no *root* in $\mathbb{Q}$. More generally, let F be any field, and let $f(x)$ be any *quadratic* polynomial. Then $f(x)$ is reducible if and only if $f(x)$ has a root in F. For if $f(x)$ has a root in F, then by Example 2 above, $f(x)$ must be reducible. Conversely, if $f(x)$ is reducible, then it must have a linear factor $g(x)$. By Example 5.6 above, $g(x)$ must have a root, say a. But if $g(a) = 0$, $f(a)$ must also be zero. (Why?)
5. Similarly, if F is any field and $f(x)$ any *cubic* polynomial in $F[x]$, then $f(x)$ is reducible if and only if $f(x)$ has a root in F. For as always, if $f(x)$ has a root, then $f(x)$ must be reducible as in Example 2 above. Conversely, if $f(x)$ is reducible, once again, as in Example 4 above, $f(x)$ must have a linear factor. (Why?) A root of this linear factor is then a root of $f(x)$.
6. Can you give an example of a fourth degree polynomial over a field F that is reducible over F but has no roots over F? Your example will thus show that Examples 4 and 5 cannot be extended to degrees higher than three.
7. If $f(x) \in F[x]$ is irreducible over F, then for any nonzero constant $c \in F$, the polynomial $cf(x)$ is also irreducible. For assume that $cf(x)$ is reducible, and let $cf(x) = g(x)h(x)$, where neither $g(x)$ nor $h(x)$ is a constant. Divide both sides by c. Why does this give you a nontrivial factorization of $f(x)$? Why is this a contradiction?

   In particular, suppose $f(x) = c_n x^n + c_{n-1}x^{n-1} + \cdots + c_0$. Write $g(x)$ for the polynomial obtained by dividing $f(x)$ by its highest coefficient, c_n, that is, $g(x) = x^n + (c_{n-1}/c_n)x^{n-1} + \cdots + (c_0/c_n)$. Then $f(x)$ is irreducible if and only if $g(x)$ is irreducible. A polynomial such as $g(x)$ whose highest coefficient is 1 is called a *monic polynomial.*

8. Every cubic polynomial in $\mathbb{R}[x]$ is reducible over $\mathbb{R}$. (See Exercise 3. Also, how is this example different from Example 5 above?) In particular, this shows that any cubic polynomial over $\mathbb{Q}$ that is irreducible over $\mathbb{Q}$ becomes reducible over $\mathbb{R}$.
9. (This is called the Fundamental Theorem of Algebra.) *Every* nonconstant polynomial $f(x) \in \mathbb{C}[x]$ has a root in $\mathbb{C}$. In other words, given any polynomial equation $f(x) = 0$ where the coefficients of $f(x)$ are complex numbers, we can find a complex number a that will solve this equation. (In particular, this applies to any polynomial equation $f(x) = 0$ whose coefficients are real numbers—since $\mathbb{R} \subseteq \mathbb{C}$. We may treat $f(x)$ as a polynomial in $\mathbb{C}[x]$, and the theorem then guarantees that we can find a complex root. Note that even though the coefficients of $f(x)$ are all real, the roots of $f(x)$ may not be real. For example, $x^2 + 1 = 0$ has the roots $\pm i$, neither of which is real.) This theorem was first proved by Gauss, who went on to give at least four different proofs. We will not prove this theorem here since it requires techniques beyond the scope of this book. (See the notes on page 149 for some remarks on the various proofs of this theorem.) It follows from this theorem and Example 2 above that if $f(x)$ is any polynomial in $\mathbb{C}[x]$ of degree greater than 1, then $f(x)$ is reducible. Described differently, the only irreducible polynomials in $\mathbb{C}[x]$ are the polynomials of degree 1.
10. The Fundamental Theorem of Algebra allows us to prove something about $\mathbb{R}[x]$ that is considerably stronger than Example 8 above. *Every polynomial in $\mathbb{R}[x]$ of degree 3 or more is reducible in $\mathbb{R}[x]$!* See Exercise 10.

Given a polynomial $f(x) \in F[x]$, how do we tell whether it is irreducible? In general, this is very difficult. Over the rationals, there are certain tests (such as Eisenstein's Criterion—see the notes on page 150)—that can be applied to *certain* polynomials to determine irreducibility. The ad hoc nature of these tests prevents them, however, from being applicable to all polynomials. There is also a lengthy algorithm due to Kronecker for finding all the factors of a polynomial with *integer* coefficients; in particular, this algorithm will determine whether the polynomial is itself irreducible. This algorithm is too laborious to be carried out by hand, but is of course amenable to computer implementation. There is also the Rational Roots Test (see Exercise 4), which can be used to study the irreducibility of cubic polynomials with integer coefficients, as in the example below:

Example 5.16
Let us use Exercise 4 to show that the polynomial $f(x) = 8x^3 - 6x - 1$ is irreducible over $\mathbb{Q}[x]$. (This is a polynomial that arises in the proof of the impossibility of the trisection of an arbitrary angle by straightedge and compass alone.) First note that since $f(x)$ is a cubic, Example 5 above shows that $f(x)$ is irreducible if and only if it has no root in $\mathbb{Q}$. So assume to the contrary that $f(x)$ is reducible. Then $f(x)$ must have a root in $\mathbb{Q}$, by Example 5. Let r/s be a root, with r and s in $\mathbb{Z}$ and $\gcd(r, s) = 1$. Note that the coefficients of $f(x)$ are *integers*. Exercise 4 hence applies, and we find that r must divide -1, and s must divide 8. Thus, the only possibilities for r are ± 1 and for s $\pm 1, \pm 2, \pm 4$, and ± 8. Described differently, the only possible roots of $f(x)$ are $\pm 1, \pm 1/2, \pm 1/4$, and $\pm 1/8$. It is a simple matter to substitute each of these eight numbers for x and verify that none of these numbers is a root of $f(x)$. Hence $f(x)$ must be irreducible!

We need to prove two facts about irreducibles before we tackle the factorization of polynomials into irreducibles.

Recall that if p is a prime number and b is any integer, then either p divides b or else p and b are relatively prime (see Lemma 1.12 of Chapter 1) . The following lemma is the analog in the context of polynomials.

Lemma 5.17
Let $p(x)$ be an irreducible polynomial in $F[x]$. Let $f(x)$ be any polynomial in $F[x]$. Then either $p(x)$ divides $f(x)$ or else $p(x)$ and $f(x)$ are relatively prime.

Proof If $p(x)$ divides $f(x)$ then we have nothing to prove, so assume that $p(x)$ does not divide $f(x)$. (This means that $f(x)$ cannot be zero—why?) What can we say about the greatest common divisor of $p(x)$ and $f(x)$? Suppose $d(x)$ is a greatest common divisor of $p(x)$ and $f(x)$.

Then $d(x)$ must divide $p(x)$, so $p(x) = d(x)h(x)$ for some polynomial $h(x)$. Since $p(x)$ is irreducible, either $d(x)$ or $h(x)$ must be a constant. We claim that $d(x)$ must be a constant. For suppose not. Then $h(x)$ must be a constant, call it k. Since k is in F, k^{-1} is also in F, and $d(x) = k^{-1}p(x)$. (Why is k nonzero?) Since $d(x)$ must also divide $f(x)$, we find that $k^{-1}p(x)$ divides $f(x)$. This implies that $p(x)$ divides $f(x)$—why? But we have assumed that $p(x)$ does not divide $f(x)$! Hence $d(x)$ must indeed be a constant. But this means that $p(x)$ and $f(x)$ are relatively prime! □

The following lemma further illustrates the analogy between irreducibles and prime numbers.

Lemma 5.18
Let $p(x)$ be an irreducible in $F[x]$. If $p(x)$ divides the product $f(x)g(x)$ of two polynomials $f(x)$ and $g(x)$ in $F[x]$, then either $p(x)$ divides $f(x)$ or else $p(x)$ divides $g(x)$.

Proof If $p(x)$ divides $f(x)$ we have nothing to prove. So assume that $p(x)$ does not divide $f(x)$. Then the previous lemma shows that $p(x)$ and $f(x)$ must be relatively prime. But by Lemma 5.13, this means that $p(x)|g(x)$. □

We are now ready for our unique prime factorization theorem for polynomials!

Theorem 5.19 (Unique Prime Factorizaton of Polynomials)
Let F be a field. Every nonconstant polynomial $f(x)$ in $F[x]$ factors into a product of irreducibles that are unique except for order and multiplication by constants.

Proof Our proof of this theorem will be very similar to our proof of the Fundamental Theorem of Arithmetic (Theorem 1.14 in Chapter 1). It would be a good idea for you to review that proof first!

We will prove the existence part first. Given the nonconstant polynomial $f(x)$, either it is irreducible or it is not. If it is irreducible, then "$f(x) = f(x)$" is its factorization into irreducibles. If it is not irreducible, then $f(x)$ must factor as $f(x) = g(x)h(x)$ for suitable polynomials $g(x)$ and $h(x)$ with $deg(g(x)) < deg(f(x))$ and $deg(h(x)) < deg(f(x))$. If both $g(x)$ and $h(x)$ are irreducible, then "$f(x) = g(x)h(x)$" is the factorization of $f(x)$ into irreducibles. If not, then either $g(x)$ or $h(x)$ must be reducible (it is quite possible, of course, that both are reducible). If, say, $g(x)$ is reducible, then $g(x) = p(x)q(x)$ for suitable polynomials $p(x)$ and $q(x)$ with $deg(p(x)) < deg(g(x))$ and $deg(q(x)) < deg(g(x))$. At this stage, we have $f(x) = g(x)h(x) = p(x)q(x)h(x)$. If all three of $p(x)$, $q(x)$, and $h(x)$ are irreducible, then "$f(x) = p(x)q(x)h(x)$" is the factorization of $f(x)$ into irreducibles. If not, then one or more of $p(x)$, $q(x)$, and $h(x)$ must be reducible....This process must eventually stop, since at each stage, the degrees of our factors are becoming smaller and smaller, and the smallest degree we are allowed to have at any stage is 1. When this process stops, we will have our factorization of $f(x)$ into irreducibles.

(Notice how we use the decrease in the *degrees* of our factors at each stage to argue that our factoring process above must stop. By contrast, in the proof of the Theorem 1.14, the *factors themselves* were decreasing.)

Now for the uniqueness. Just as in the proof of the uniqueness part of Theorem1.14, the key is to recognize that if a nonconstant polynomial $f(x)$ has two factorizations into irreducibles, then some irreducible in the first factorization must equal some irreducible in the second factorization *except for multiplication by some constant*. (This is the chief difference between the proof of uniqueness part of this theorem and the proof of the uniqueness part of Theorem 1.14.) We will then cancel irreducibles pair by pair in the two factorizations, and conclude that the two factorizations must be the same.

Given the nonconstant polynomial $f(x)$, assume that we have the two factorizations $f(x) = p_1(x) \cdots p_s(x) = q_1(x) \cdots q_t(x)$ into irreducibles. We may assume without any loss of generality that $s \geq t$, and we may further assume that $s \neq 1$. (For if $s = 1$, then since $s \geq t$, $t = 1$ as well. The two factorizations of $f(x)$ will therefore be $f(x) = p_1(x)$ and $f(x) = q_1(x)$, from which we can conclude immediately that both $p_1(x)$ and $q_1(x)$ have to be the same irreducible, since they both equal $f(x)$. In other words, the two factorizations would be the same, and there would be nothing to prove!)

Since $p_1(x)$ divides a, and since $a = q_1(x) \cdots q_t(x)$, $p_1(x)$ must divide $q_1(x) \cdots q_t(x)$. By Exercise 2 (which generalizes Theorem 5.18), $p_1(x)$ divides one of the irreducibles $q_i(x)$, and by relabeling the q_i if necessary, we may assume that $p_1(x)$ divides

$q_1(x)$. Thus $q_1(x) = p_1(x)c_1(x)$ for some polynomial $c_1(x)$ in $F[x]$. Since $q_1(x)$ is irreducible either $p_1(x)$ or $c_1(x)$ must be a constant, and since $p_1(x)$ cannot be constant (why not?), $c_1(x)$ must be a constant! Writing c_1 instead of $c_1(x)$ to emphasize that it is a constant, we find that $q_1(x) = c_1p_1(x)$, that is, the two irreducibles $q_1(x)$ and $p_1(x)$ are the same except for multiplication by a constant.

Now write $p_1(x)$ as $c_1^{-1}c_1p_1(x)$, and cancel $c_1p_1(x)$ from the first factorization and $q_1(x)$ from the second factorization. (Remember, $F[x]$ is an integral domain. As we saw in Lemma 2.19, we can always cancel nonzero elements from both sides of an equation.) Because of our assumption that $s \neq 1$ we must have $s \geq 2$. Thus, there is at least one more factor $p_2(x)$ in the first factorization of $f(x)$. Let us fold c_1^{-1} into $p_2(x)$, and call their product $p_2'(x)$. Then $p_2'(x)$ is also irreducible. (Why?) We thus have $p_2'(x)\cdots p_s(x) = q_2(x)\cdots q_t(x)$. If there are no irreducibles on the right side of this equation (this would happen if $t = 1$), then we would find that the product of irreducibles $p_2'(x)\cdots p_s(x)$ equals 1, which is absurd. Hence, $t \geq 2$. Now we repeat our arguments above with $p_2'(x)$, and relabeling the $q_i(x)$ if necessary, we show that except for multiplication by a constant, $p_2'(x)$ must be the same as $q_2(x)$. Since $p_2'(x)$ is the same as $p_2(x)$ except for multiplication by a constant, we find that $p_2(x)$ is the same as $q_2(x)$ except for multiplication by a constant. Proceeding similarly, we can see that $s = t$, and after relabeling if necessary, each $p_i(x)$ ($i = 1, \ldots, s$) must be the same as $q_i(x)$ except for multiplication by a constant. □

(At this point, you may want to read about the similarities between $\mathbb{Z}$ and $F[x]$ in the notes on page 150.)

We will finish this chapter with a discussion on the number of roots of a polynomial.

We saw ealier in the chapter (see the paragraph after Definition 5.5) that $a \in F$ is a root of the nonzero polynomial $f(x) \in F[x]$ if and only if $(x-a)|f(x)$. Now, if a is a root, it is quite possible that not only does $(x-a)$ divide $f(x)$, but that a higher power of $(x-a)$ also divides $f(x)$. For example, 1 is a root of the polynomial $x^3 - 4x^2 - 3x - 2$, but $(x-1)^2$ divides this polynomial, since $x^3 - 4x^2 - 3x - 2$ factors in $\mathbb{Q}[x]$ as $(x-1)^2(x-2)$. This leads to the following:

Definition 5.20
Let F be a field, and let $f(x) \in F[x]$ be a nonconstant polynomial. We say that an element $a \in F$ is a root *of multiplicity* k of $f(x)$ if $(x - a)^k$ divides $f(x)$ but $(x - a)^{k+1}$ does not divide $f(x)$.

(So, in the example above, 1 is a root of multiplicity 2 of the polynomial $x^3 - 4x^2 - 3x - 2$, while 2 is a root of multiplicity 1.)

Now, given some nonconstant polynomial $f(x) \in F[x]$, F a field, suppose that $a_1, a_2, \ldots, a_t$ in F are distinct roots of $f(x)$, of respective multiplicities $k_1, k_2, \ldots, k_t$. Consider the factorization of $f(x)$ into irreducibles of $F[x]$, say $f(x) = p_1(x) \cdot p_2(x) \cdot \ldots \cdot p_r(x)$. Since $(x - a_1)$ divides this product of irreducibles, $(x - a_1)$ must divide some irreducible, say $p_1(x)$, from which it follows that $p_1(x) = c_1(x - a_1)$ for some constant c_1. (Why? You have seen the relevant arguments in the proof of Theorem 5.19 above.) Thus, $(x - a_1)$ is (up to multiplication by c_1) one of the irreducible factors of $f(x)$. But a_1 is a root of multiplicity k_1, so $(x - a_1)^{k_1}$ divides $f(x)$. Thus, $(x - a_1)$ must appear k_1 times in the factorization of $f(x)$ into irreducibles. Applying the same reasoning to $(x - a_2)$, we find that $(x - a_2)$ must be, up to multiplication by some constant c_2, one of the irreducible factors of $f(x)$. Since $a_1 \neq a_2$ (the a_i are distinct), $(x - a_2)$ must be a *different* irreducible factor of $f(x)$ from $(x - a_1)$. Moreover, $(x - a_2)$ must appear k_2 times in the factorization of $f(x)$. Proceeding similarly, we find that up to multiplication by some constant c_i, each of the linear polynomials $(x - a_i)$ $(i = 1, \ldots, t)$ must be part of the irreducible factorization of $f(x)$, and that each $(x - a_i)$ must appear k_i times. Collecting all these constants c_i as well as the remaining irreducible factors of $f(x)$ together and calling their product $g(x)$, we find that we have proved the following:

Lemma 5.21
Let F a field and $f(x) \in F[x]$ a nonconstant polynomial. If $f(x)$ has distinct roots $a_1, a_2, \ldots, a_t$ in F of respective multiplicities $k_1, k_2, \ldots, k_t$, then $f(x) = g(x)(x - a_1)^{k_1}(x - a_2)^{k_2} \ldots (x - a_t)^{k_t}$ for some polynomial $g(x) \in F[x]$.

Now, there is an immediate conclusion that we can draw from this lemma: the number of roots of a nonconstant polynomial $f(x) \in F[x]$ in F, counting each root as often as its multiplicity, cannot exceed the degree of the polynomial! (Simply note that if $f(x) = g(x)(x-a_1)^{k_1}(x-a_2)^{k_2}\ldots(x-a_t)^{k_t}$, then the degree of $f(x)$ must equal the sum of the degree of $g(x)$ and the degrees of the factors $(x-a_i)^{k_i}$, from which the conclusion immediately follows.) We will state this as a separate corollary:

Corollary 5.22
Let F be a field, and let $f(x) \in F[x]$ be a nonconstant polynomial. Then the number of roots of $f(x)$ in F, counting each root as often as its multiplicity, cannot exceed the degree of $f(x)$.

It must be emphasized that the considerations above applied to the roots of the polynomial $f(x) \in F[x]$ that were in F itself. Of course, over a given field F, a polynomial $f(x) \in F[x]$ need not have a root in F at all—we have seen lots of examples of this already. However, if we allow our field to be $\mathbb{C}$, then the Fundamental Theorem of Algebra (see Example 5.15.9) guarantees that every nonconstant polynomial has a root, and we can use this theorem to say something more about the roots of $f(x)$—the number of roots of $f(x)$ in $\mathbb{C}$, counting each root as often as its multiplicity, must *equal* the degree of $f(x)$! To prove this, note that since the only irreducibles in $\mathbb{C}[x]$ are of degree 1, the factorization of our nonconstant polynomial $f(x) \in \mathbb{C}[x]$ into irreducibles must consist only of linear factors. Since any linear polynomial $ax + b$ can be rewritten as $a(x - (-b/a))$, every linear factor of $f(x)$ is of the form $c(x - a)$ for some constants c and a. Collecting all these constants c that come from the various factors together into a single constant c, we find that $f(x)$ factors as this constant c times various linear factors of the form $(x - a_1)$, $(x - a_2)$, etc., for various complex numbers a_1, a_2, etc. If each factor $(x-a_i)$ appears k_i times, and if there are t such different linear factors, we find that $f(x)$ factors as $c(x-a_1)^{k_1}(x-a_2)^{k_2}\ldots(x-a_t)^{k_t}$. By the very definition of a root of multiplicity k, each of these a_i must be a root of $f(x)$ of multiplicity k_i. By comparing degrees, we find that $\sum_{i=0}^{t} k_i = n$, where $n = deg(f)$. In other words, the total number of these particular roots a_i, counting multiplicities, already equals n. It follows that these a_i are *all* the roots of $f(x)$, since by Corollary 5.22 above, the number of roots of f, counting multiplicities, cannot exceed n. Moreover, we can immediately describe the constant c:

if we compare the highest terms on both sides of the factorization $f(x) = c(x-a_1)^{k_1}(x-a_2)^{k_2}\ldots(x-a_t)^{k_t}$, we find that c must equal the highest coefficient of f! We have thus proved the following:

Theorem 5.23
Let $f(x) \in \mathbb{C}[x]$ be a nonconstant polynomial. Then the number of roots of $f(x)$ in $\mathbb{C}$, counting each root as often as its multiplicity, equals the degree of $f(x)$. If the various roots of $f(x)$ are $a_1, a_2, \ldots, a_t$, of respective multiplicities $k_1, k_2, \ldots, k_t$, then $f(x)$ factors in $\mathbb{C}[x]$ as $f(x) = c(x - a_1)^{k_1}(x-a_2)^{k_2}\ldots(x-a_t)^{k_t}$, where c equals the highest coefficient of $f(x)$.

This theorem applies, in particular, to any polynomial $f(x)$ whose coefficients lie in some subfield F of $\mathbb{C}$—simply view $f(x)$ as a polynomial in $\mathbb{C}[x]$! Thus, while $f(x)$ may not have even a single root in the original field F, this theorem guarantees that over the complex numbers, $f(x)$ will have as many roots as its degree.

Exercises

1. Prove Lemma 5.3.
2. Use induction on n and the basic ideas of the proof of Lemma 5.18 to prove the following generalization—if $p(x)$ in $F[x]$ is irreducible and if $p(x)$ divides the product $f_1(x)\cdots f_n(x)$ (where the $f_i(x)$ are in $F[x]$ and n is some positive integer) then $p(x)$ divides some $f_i(x)$.
3. This exercise is designed to prove that all polynomials in $\mathbb{R}[x]$ of *odd* degree, other than the linear polynomials, are reducible.

   (a) Let $f(x) = \sum_{i=0}^{n} a_i x^i$ be a polynomial in $\mathbb{R}[x]$. Viewing $f(x)$ as a continuous *function* from $\mathbb{R}$ to $\mathbb{R}$, prove that
   $$\lim_{x\to\infty}\frac{f(x)}{a_n x^n} = \lim_{x\to-\infty}\frac{f(x)}{a_n x^n} = 1.$$

   (b) Part (a) shows that the ratio of $f(x)$ and $a_n x^n$ can be made arbitrarily close to 1 by taking x to be a suitably large positive number or a suitably large negative number. In more precise language, given any positive real number ϵ, no matter how small, there exists a positive integer N such that for all $x > N$ and for all $x < -N$, $1 - \epsilon < \frac{f(x)}{a_n x^n} < 1 + \epsilon$. Now assume that n is odd. Study the behavior of $a_n x^n$ as $x \to \pm\infty$ to conclude that either 1) $f(x)$ is negative when x is a suitably large negative number and positive when x is a suitably large positive number, or 2) $f(x)$ is positive when x is a suitably large negative number and negative when x is a suitably large positive number.

(c) Conclude using the Intermediate Value Theorem that when n is odd, the graph of $f(x)$ must cross the x-axis.

(d) Conclude that if $f(x) \in \mathbb{R}[x]$ is of odd degree and if $deg(f(x)) \neq 1$, then $f(x)$ is reducible.

4. (Rational Roots Test.) Suppose $f(x)$ is a nonconstant polynomial with integer coefficients, that is, $f(x) = a_n x^n + a_{n-1}x^{n-1} + \cdots + a_1 x + a_0$, where $n > 0$ and the a_i are integers. Now suppose that $f(x)$ has a root in $\mathbb{Q}$, that is, $f(c) = 0$ for some $c \in \mathbb{Q}$. Write c as r/s, where r and s are integers with $\gcd(r, s) = 1$. Prove that s must divide a_n and r must divide a_0. Now show that if $a_n = 1$, then c must be an integer. (Hint: Multiply the equation $f(r/s) = 0$ by s^n, move first $a_n r^n$ and then $a_0 s^n$ to one side, and stare hard at the results.)

5. We will study in this exercise how the coefficients of a polynomial depend on its roots.

   (a) Suppose the quadratic polynomial $f(x) = ax^2 + bx + c$ has the two roots r_1 and r_2 in $\mathbb{C}$. (Since f is assumed quadratic, $a \neq 0$.) Using Theorem 5.23, show that $r_1 + r_2 = -b/a$ and $r_1 r_2 = c/a$.

   (b) Suppose that the cubic polynomial $f(x) = ax^3 + bx^2 + cx + d$ has the three roots r_1, r_2, and r_3 in $\mathbb{C}$. Show that $r_1 + r_2 + r_3 = -b/a$, $r_1 r_2 + r_1 r_3 + r_2 r_3 = c/a$, and $r_1 r_2 r_3 = -d/a$.

   (c) Now suppose that the nth degree polynomial $f(x) = a_n x^n + a_{n-1}x^{n-1} + \cdots + a_1 x + a_0$ has the n roots $r_1, r_2, \ldots, r_n$ in $\mathbb{C}$. What would be the statements analogous to those in parts 5a and 5b that you ought to prove? Now prove them!

What this exercise shows is that given all the roots of a *monic* polynomial, we can recover the coefficients of the polynomial from our knowledge of the roots—simply apply the results of this exercise with $a_n = 1$. The expressions for $-a_{n-1}, a_{n-2}, -a_{n-3}, \ldots, (-1)^n a_0$ that you will get in part 5c are known as the *elementary symmetric functions in n variables.* For instance, the elementary symmetric functions in two variables x_1 and x_2 (see part 5a) are $x_1 + x_2$ and x_1x_2. The elementary symmetric functions in three variables x_1, x_2, and x_3 (see part 5b) are $x_1 + x_2 + x_3$, $x_1x_2 + x_1x_3 + x_2x_3$, and $x_1x_2x_3$.)

6. We saw the definitions of the roots of unity ω_n in Exercise 10 of Chapter 4. We will see in this exercise how these roots of unity are used to determine nth roots of complex numbers.

   We know that every nonzero real number has two distinct square roots, and one square root is just the negative of the other, that is, one square root is just -1 times the other. Now -1 is just ω_2, so another way of phrasing the statement above is that every nonzero real number has two distinct square roots, and if a is one square root, then $\omega_2 a$ is the other. Generalizing this, show that for any integer $n \geq 2$, every nonzero *complex* number c has n distinct nth roots, and if a is any one nth root, then $\omega_n a$, $\omega_n^2 a$, $\ldots$, $\omega_n^{n-1} a$ are the remaining $n-1$ nth roots. (Hint: Consider the equation $x^n - c = 0$. This equation has at least one solution in $\mathbb{C}$—why? Call this solution a. Show that $\omega_n a$, $\omega_n^2 a$, $\ldots$, $\omega_n^{n-1} a$ are also solutions. Use Exercise 10c of Chapter 4 to show that these solutions are all distinct, and use Theorem 5.23 above to show that there can be no other solutions.)

   In particular, taking c to be a real number, this exercise shows that just as every nonzero real number has two distinct square roots, every nonzero real number also has *three* distinct cube roots, *four* distinct fourth roots, and so on. Of course, these various cube roots, fourth roots, etc., will not themselves be real numbers in general. (For instance, the three cube roots of 8 are 2, $2(\cos(2\pi/3) + i\sin(2\pi/3))$, and $2(\cos(4\pi/3) + i\sin(4\pi/3))$, of which only 2 is a real number.)

7. Let F be any field, and suppose $f(x)$ and $g(x)$ are two nonzero polynomials in $F[x]$ of degree n such that $f(a_i) = g(a_i)$ for distinct

elements $a_i \in F$, $i = 0, 1, \ldots, n$. Prove that $f(x)$ must equal $g(x)$. (Hint: Proving $f(x) = g(x)$ is the same as proving $f(x) - g(x) = 0$. What ideas do you get from Corollary 5.22 above?)

8. Suppose $a_0, a_1, \ldots, a_n$ are distinct elements of F, and suppose $b_0, b_1, \ldots, b_n$ are arbitrary elements of F. Prove that the polynomial

$$f(x) = \sum_{i=0}^{n} \frac{b_i \prod_{j=0, j \neq i}^{n} (x - a_j)}{\prod_{j=0, j \neq i}^{n} (a_i - a_j)}$$

is the *unique* polynomial of degree n such that $f(a_i) = b_i$, $i = 0, 1, \ldots, n$. (Hint: Exercise 7 above!)

This exercise finds use in the following: Suppose you have $n + 1$ data points on a graph: (a_0, b_0), (a_1, b_1), ..., (a_n, b_n), and suppose you wish to find an nth degree polynomial that passes through all these $n + 1$ data points. The formula above would give you precisely the polynomial that you would be seeking! This formula is known as *Lagrange's Interpolation Formula*.

9. Let K/F be a field extension.

   (a) We have seen several examples in the text of polynomials $f(x) \in F[x]$ that are irreducible over F but become reducible over K. This exercise goes the other way: Prove that if a polynomial $f(x) \in F[x]$ is irreducible over K, then it must be irreducible over F as well.

   (b) This exercise shows that in contrast to irreducibility, relative primeness is preserved when you go from F to K: Suppose that $f(x)$ and $g(x)$ are two polynomials in $F[x]$ that are relatively prime in $F[x]$. Show that they remain relatively prime as polynomials in $K[x]$ as well. (Hint: Corollary 5.12.)

   (c) This exercise shows that given two polynomials in $F[x]$, if one divides the other when considered as polynomials in $K[x]$, then the one already divides the other as polynomials in $F[x]$: Prove that if $f(x)$ and $g(x)$ are in $F[x]$, and if $f(x) = g(x)h(x)$ for some $h(x)$ in $K[x]$, then $h(x)$ *has its coefficients in* F, that is, $h(x) \in F[x]$. (Hint: Apply the division algorithm to $f(x)$ and

$g(x)$, first in $F[x]$ and then in $K[x]$, and invoke the fact that the quotient and remainder are unique.)

10. In Exercise 3 above, we saw that all polynomials in $\mathbb{R}[x]$ of odd degree, other than the linear polynomials, are reducible. The proof invoked notions of continuity and limits from calculus, but did not invoke any knowledge of the complex numbers, and in particular, did not assume the Fundamental Theorem of Algebra (Example 5.15.9). However, if we assume the Fundamental Theorem of Algebra, we can actually prove something stronger: *all polynomials of degree* 3 *or higher in* $\mathbb{R}[x]$ *are reducible over* $\mathbb{R}$*!* (In other words, among the polynomials of degree three or more, it is not just the odd-degree ones that are reducible, but as well, polynomials of degree 4, 6, etc.) The proof is in several steps:

    (a) Recall that if $z = a + ib$ is a complex number, then the complex conjugate of z, denoted $\overline{z}$, is defined by $\overline{z} = a - ib$. Prove that $\overline{z_1 + z_2} = \overline{z_1} + \overline{z_2}$ and $\overline{z_1 z_2} = \overline{z_1}\,\overline{z_2}$.

    (b) Now prove using induction that if $f(z) = a_n z^n + a_{n-1} z^{n-1} + \cdots + a_0$ is a polynomial expression in z with complex coefficients, then $\overline{f(z)} = \overline{a_n}(\overline{z})^n + \overline{a_{n-1}}(\overline{z})^{n-1} + \cdots + \overline{a_0}$.

    (c) Let $f(x)$ be a polynomial with *real* coefficients. Since $f(x)$ will automatically be irreducible if *deg*$(f(x)) = 1$, let us assume that *deg*$(f(x)) \geq 2$. If $f(x)$ has a *real* root α, then $f(x)$ is automatically reducible in $\mathbb{R}$ (why?). So assume that $f(x)$ does not have a root in $\mathbb{R}$. By the Fundamental Theorem of Algebra, $f(x)$ must have a complex root $z = a + ib$, where $b \neq 0$ (because of our assumption that $f(x)$ has no real root). Use part 10b above and the fact that the coefficients of $f(x)$ are real to prove that $\overline{z}$ is also a root of $f(x)$. This is described as follows: *complex nonreal roots of polynomials with real coefficients come in conjugate pairs.*

    (d) Use Theorem 5.23 to show that $(x - z)(x - \overline{z})$ divides $f(x)$ as polynomials in $\mathbb{C}[x]$.

    (e) Show that if x is any complex number, then $x + \overline{x}$ and $x\overline{x}$ are both *real* numbers. (Note the similarity with Exercise 7(a)i of Chapter 4. If you are interested, formulate the analog of Exercise 7(a)ii of that same chapter and prove it too!)

(f) Use part 10e above to conclude that the coefficients of the polynomial $(x - z)(x - \overline{z})$ are *real.*

(g) Now conclude using parts 10d and 10f and Exercise 9c above that if $deg(f(x)) \geq 3$, then $f(x)$ is reducible over $\mathbb{R}$.

Note that the existence of a root $z \in \mathbb{C}$ was crucial to the proof!

11. This exercise explains how synthetic division works! F, as usual, is a field. Working in $F[x]$, suppose we wish to divide the non-constant polynomial $f(x)$ by $x - c$, where c is any constant. Thus, we wish to find a polynomial $q(x)$ and a constant r such that $f(x) = q(x)(x - c) + r$. (Why is the remainder a constant?) Let $f(x) = f_0x^n + f_1x^{n-1} + \cdots f_{n-1}x + f_n$. (Since $f(x)$ is not a constant, $n \geq 1$. Note that we have reversed the indices, preferring to call the highest-degree coefficient f_0 rather than f_n—this is just for convenience.)

(a) Prove that $g(x)$ must be of degree $n - 1$.

(b) So assume that $g(x) = g_0x^{n-1} + g_1x^{n-2} + \cdots + g_{n-2}x + g_{n-1}$. Prove that $g_0 = f_0$, $g_1 = f_1 + cg_0$, $g_2 = f_2 + cg_1$, …, $g_{n-1} = f_{n-1} + cg_{n-2}$, $r = f_n + cg_{n-1}$. Thus, we can recursively solve for the g_i: the first equation gives us g_0, once we determine g_0, the second equation gives g_1, etc.

This result yields the following algorithm for determining the quotient $q(x)$ and reminder r:

$$
\begin{array}{c|cccccc}
c\,) & f_0 & f_1 & f_2 & \cdots & f_{n-1} & f_n \\
 & & cg_0 & cg_1 & \cdots & cg_{n-2} & cg_{n-1} \\
+ & & & & & & \\
\hline
 & g_0 & g_1 & g_2 & \cdots & g_{n-1} & r
\end{array}
$$

12. We will solve some concrete equations in this problem!

(a) Solve $x^4 + x^3 - x^2 - 2x - 2 = 0$ if you know that $\pm\sqrt{2}$ are two of its roots.

(b) Solve $x^5 - 2x^4 + 2x^3 - 2x^2 + 4x - 4 = 0$ if you know that $1 + i$ is one root. (Hint: See Exercise 10c above. You will also find Exercise 6 above useful.)

(c) Solve $x + \dfrac{1}{x} = y$ for x in terms of y.

(d) Solve $x^4 + x^3 + x^2 + x + 1 = 0$ by first dividing through by x^2, making the substitution $y = x + \frac{1}{x}$, and then deriving a quadratic equation for y. (Where have you seen the polynomial $x^4 + x^3 + x^2 + x + 1$ before?) Notice that this technique could be used for any quartic equation of the form $ax^4 + bx^3 + cx^2 + bx + a$.

(e) Show that if $x + y = a$ and $xy = b$, then x and y are both roots of the quadratic polynomial $t^2 - at + b = 0$.

(f) Follow the notes on Cardano's solution of the cubic equation on page 153 and solve the equation $x^3 + 3x + 2$.

Now that you have solved some polynomial equations yourselves, you may find the remarks on page 152 on the solution of polynomial equations interesting!

Notes

Remarks on the division algorithm for polynomials over arbitrary rings Notice something about the long division process and the uniqueness arguments following Equation 5.4. There are only two places where we invoke the fact that F is a field. The first is in the long division, when we divide $g(x)$ by its highest coefficient. (This happens both at the beginning of the division process, where we multiply $g(x)$ by a suitable constant times a suitable power of x before subtracting the result from $f(x)$, and at subsequent steps in the process, where we again multiply $g(x)$ by a suitable constant times a suitable power of x, this time before subtracting the result from the polynomial obtained from the *previous* step.) The second place is in the arguments after Equation 5.4, where we invoke the fact that the degree of the product of a nonzero polynomial and $g(x)$ must be greater than or equal to the degree of $g(x)$.

As far as the long division process is concerned, observe that *the only element in F that we need to be able to divide by is the highest coefficient of $g(x)$!* This shows that the long division process will actually work for polynomials with coefficients in *any* ring R, as long as we assume that the highest coefficient of $g(x)$ is invertible in the ring! (Recall from Exercise

9 in Chapter 2 that an element a in a ring R is said to be *invertible* if there exists another element b in R such that $ab = ba = 1$.)

As for the uniqueness arguments following Equation 5.4, try to prove for yourselves using the ideas described in the chapter notes to Example 2.7.7 in Chapter 2 as well as the statement of Exercise 9 of the same chapter the following fact: if the highest coefficient of $g(x)$ is invertible in R, then $deg(h(x)g(x)) = deg(h(x)) + deg(g(x))$ for any nonzero $h(x) \in R[x]$. This shows that as long as the highest coefficient of $g(x)$ is invertible, the uniqueness arguments that followed Equation 5.4 will continue to work for polynomials from an arbitrary ring, and the quotient and remainder we obtain from the long division process will be unique.

We state this below as a theorem.

Theorem 5.24 (Division Algorithm)
Let R be any ring and $f(x)$ and $g(x)$ be in $R[x]$, with $g(x) \neq 0$. Assume that the highest coefficient of $g(x)$ is invertible in R. Then there exist unique polynomials $q(x)$ and $r(x)$ in $F[x]$ with either $r(x) = 0$ or $0 \leq deg(r(x)) < deg(g(x))$, such that $f(x) = q(x)g(x) + r(x)$.

As an example, dividing $f(x) = x^3 + x + 1$ by $g(x) = x - 1$ (notice that both $f(x)$ and $g(x)$ are in $\mathbb{Z}[x]$ and the highest coefficient of $g(x)$ is 1, which is invertible in $\mathbb{Z}$), we find that $x^3 + x + 1 = (x^2 + x + 2)(x - 1) + 3$, so that $q(x) = x^2 + x + 2$ and $r(x) = 3$. Notice that both $q(x)$ and $r(x)$ also have coefficients in $\mathbb{Z}$.

Remarks on the notion of one polynomial being larger than another It is worth pointing out here that although the degree of a polynomial is a good enough notion of size to yield a factorization theorem for polynomials (see the remarks on page 150 on the similarities between $\mathbb{Z}$ and $F[x]$), it is not good enough to measure whether one polynomial is larger than another or not. The problem is that two polynomials can have the same degree, yet not be equal to each other, not even up to multiplication by a constant. For those of you who know about order relations on sets, the problem, at a more conceptual level, is that the degree of a polynomial does not provide a *total order* on the set of polynomials over a field.

Remarks on the definition of the greatest common divisor Although the degree of a polynomial does not provide a total order on the set of polynomials over a field (see the remarks immediately above), it is possible to use the degree to provide an alternative definition of the greatest common divisor. One can define the greatest common divisor of two nonzero polynomials $f(x)$ and $g(x)$ as a common divisor of largest possible degree. Notice that any common divisor of $f(x)$ and $g(x)$ cannot have degree larger than $min\{deg(f(x)), deg(g(x))\}$ (why?), so it certainly makes sense to talk of a common divisor of largest degree. The problem is that of uniqueness: there could be several common divisors of largest possible degree. Why should any two such common divisors be the same up to multiplication by constants? This can of course be proved (using the ideas behind Theorem 5.10). The reason we have chosen to define the greatest common divisor as in Definition 5.9 is to allow you to become familiar with a definition of greatest common divisor that applies to all commutative rings, something you will hopefully find useful in your future studies!

Remarks on the Fundamental Theorem of Algebra (Example 5.15.9) All known proofs of the Fundamental Theorem of Algebra ultimately depend on theorems derived from the topology of the complex plane, that is, on theorems that use the notion of "nearness" inherent to the complex numbers. An example of such a theorem is *Liouville's Theorem*, which says that a bounded analytic function (from the complexes to the complexes) must be a constant. The notion of an analytic function is based on the notion of one complex number *approaching* another, or in other words, coming arbitrarily *near* another. Liouville's Theorem is thus ultimately based on the topology of the complexes. This theorem perhaps affords one of the shortest proofs of the Fundamental Theorem of Algebra: If a nonconstant polynomial $f(x) \in \mathbb{C}[x]$ has no root, then the function $g(x) = 1/f(x)$ is easily seen to be bounded and analytic, and hence by this theorem, a constant. But this says that $f(x)$ is also a constant function, which is a clear contradiction!

Exercise 3 leads to another proof of the Fundamental Theorem of Algebra. (Notice that the statements in Exercise 3 use notions like continuity and limits, which are notions based on the *nearness* of two real numbers, that is, on the topology of the reals. The topology of the reals is just that induced by the topolgy of the complexes, so we have once again invoked the topology of the complexes!) This proof, once the statements in Exercise 3 have been proved, is quite algebraic, and involves *Galois*

Theory, which is a theory of the solutions of polynomial equations, and in particular, of the solvability of polynomial equations by radicals (see the remarks on this topic on page 152).

Most introductory books on complex analysis contain at least one (if not several!) proofs of the Fundamental Theorem of Algebra.

Remarks on Eisenstein's Irreducibility Criterion The criterion is as follows: Let $f(x) = a_n x^n + a_{n-1}x^{n-1} + \cdots + a_1 x + a_0$ be a polynomial with *integer* coefficients. Suppose the a_i have the property that for some prime p, $p \nmid a_n$, $p|a_{n-1}$, $p|a_{n-2}$, ..., $p|a_0$, but $p^2 \nmid a_0$. Then $f(x)$ is irreducible as a polynomial in $\mathbb{Q}[x]$.

This is not too hard to prove, but the proof first involves proving Gauss's Lemma, which states that if a polynomial in $\mathbb{Z}[x]$ factors into a product of two polynomials with coefficients in $\mathbb{Q}$, then it also factors into a product of two polynomials with coefficients in $\mathbb{Z}$. Since this is carrying us a little too far afield, we will omit the proof.

As an example, Eisenstein's Irreducibility Criterion shows that $3x^4 - 2x^2 + 6$ is irreducible over $\mathbb{Q}$ (take $p = 2$). Similarly, Eisenstein's Criterion shows that the polynomials $x^n - p$ ($n = 1, 2, \ldots$) are all irreducible over $\mathbb{Q}$ for any prime p.

Eisenstein's Irreducibility Criterion is crucial in proving that for any prime p, the polynomial $x^{p-1} + x^{p-2} + \cdots + x + 1$ is irreducible over $\mathbb{Q}$. (The proof is indirect: Write $f(x)$ for the given polynomial. One first shows that $f(x)$ is irreducible if and only if $f(x+1)$ is irreducible, and one then applies Eisenstein's Criterion to $f(x+1)$.) The polynomial $x^{p-1} + x^{p-2} + \cdots + x + 1$ is known as the *pth cyclotomic polynomial*. The fact that it is irreducible is crucial to the analysis of the constructibility of regular p-gons. (You have seen polynomials of this form before, in Exercise 10 of Chapter 4.)

Remarks on the similarities between $\mathbb{Z}$ and $F[x]$ If one were to study the development of unique prime factorization in $\mathbb{Z}$ in Chapter 1 and the development of unique prime factorization in $F[x]$ in this chapter in parallel, one would notice some remarkable similarities. (And no doubt, you have already noticed these similarities yourselves!) If one were to examine these similarities more closely, one would notice that the central feature of both rings is the existence of just the right notion of "size" that leads to a reasonable division algorithm. In turn, it is the existence of a division algorithm that leads to the crucial result that the greatest common divisor is expressible as a linear combination (Theorems 1.6

and 5.10), which then allows us to say that if a prime integer (Lemma 1.13) or an irreducible polynomial (Lemma 5.18) divides a product, then it must divide one of the factors of the product. This last result is crucial to proving the uniqueness of prime factorization—it was because of this result that we were able to cancel primes and irreducibles pair by pair and conclude that any two factorizations of an integer or polynomial must be the same. The "size" is useful in another context as well, namely, for proving the existence of prime factorization in $\mathbb{Z}$ and factorization into irreducibles in $F[x]$.

So what is this "size" in the two rings? For $\mathbb{Z}$, the size of an integer is just the absolute value of the integer, and for $F[x]$, the size of a polynomial is just its degree! Notice how the division algorithm specifies the remainder r on dividing an integer a by a positive integer b as either being zero or of smaller absolute value than b. (Since r and b are nonnegative, to say $r = 0$ or $0 \leq r < b$ is really the same as saying that $r = 0$ or $0 \leq |r| < |b|$.) Similarly, notice how the division algorithm specifies the remainder on dividing a polynomial $f(x)$ by a nonzero polynomial $g(x)$ as either being zero or of degree smaller than $g(x)$. Also, notice how we determined the gcd(a, b) for two integers a and b as the element of smallest absolute value in the set P of Theorem 1.6, and determined gcd$(f(x), g(x))$ as an element of least degree in the set S of Theorem 5.10. Finally, notice how we obtained the existence of the prime factorization of an integer a by successively factoring it into products of integers of smaller absolute value, and how we obtained the existence of the factorization into irreducibles of the polynomial $f(x)$ by successively factoring it into products of polynomials of smaller degree.

Mathematicians have studied these two seemingly different notions of size and have wondered about the existence of a similar notion of size in other rings. They have wondered what property such a size should have if it were to lead to the existence and uniqueness of factorization into irreducibles in other rings beside $\mathbb{Z}$ and $F[x]$. (Recall that we have defined the concept of an irreducible element for arbitrary commutative rings in Exercise 10 of Chapter 2.) This question is well understood now; it is known that the size should satisfy the following properties (here, R is an arbitrary integral domain):

1. The size is a function d from $R - \{0\}$ to $\mathbb{N}$, that is, for any nonzero element $a \in R$, $d(a)$ is a nonnegative integer.

2. For a and b in $R - \{0\}$, $d(a) \leq d(ab)$.

3. For a and b in R, $b \neq 0$, there exist q and r in R such that $a = bq + r$, where either $r = 0$ or else $d(r) < d(b)$.

An integral domain that has such a size function is known as a *Euclidean domain.* It can be proved that *every noninvertible element of a Euclidean domain factors into a product of irreducibles, which are unique in a suitable sense!*

What are some examples of Euclidean domains? As you would expect, both $\mathbb{Z}$ and $F[x]$ are Euclidean domains. So is any field F (simply take $d(a) = 1$ for every nonzero a). But then there are other examples as well. Perhaps the most familiar one is the ring of Gaussian integers $\mathbb{Z}[i]$. The size of a nonzero element here is given by $d(a + ib) = a^2 + b^2$. The fact that $\mathbb{Z}[i]$ is a Euclidean domain (and hence has the property of unique factorization of noninvertible elements into irreducibles) is useful in proving, for instance, that every prime integer of the form $4n + 1$ is expressible as $a^2 + b^2$ for suitable integers a and b. This alone will hopefully convince you of the worth of a more abstract approach to mathematics!

Other examples of Euclidean domains are $\mathbb{Z}[\sqrt{-2}]$, $\mathbb{Z}[\sqrt{2}]$, and $\mathbb{Z}[\sqrt{3}]$. (Do not assume from this list that $\mathbb{Z}[\sqrt{d}]$ is a Euclidean domain for all integers d. That is false!)

Remarks on solutions of polynomial equations Suppose that we have a polynomial equation $f(x) = 0$ that we wish to solve, where the coefficients of f are in some subfield of the complex numbers (for instance, in the rationals). We have seen in Theorem 5.23 that f will have as many roots in $\mathbb{C}$ (counting multiplicities) as its degree. The following question arises: is there a formula involving sums, products, and various mth roots that describes these roots in terms of the coefficients of the polynomial $f(x)$? What we are after is something analogous to the quadratic formula, which describes the roots of the quadratic equation $ax^2 + bx + c = 0$ as $\dfrac{-b \pm \sqrt{b^2 - 4ac}}{2a}$. When such a formula exists, we say our equation is *solvable by radicals.*

There certainly are such formulas for cubic equations (see the remarks below) and quartic (degree 4) equations, and these formulas have been known since about the sixteenth century, but for the longest time, it was unknown whether solutions of fifth degree equations could be expressed by such a formula. It was proved by Abel in the early part of the nineteenth century, however, that no such formula exists for a general fifth degree equation! (A proof of this result had been furnished slightly earlier by Ruffini, but that proof was incomplete.) But more: a whole theory of solvability by radicals (now known as *Galois theory*) was created by Galois. This theory describes whether an equation can be solved by radicals or not in terms of a certain *finite* group, and is considered by many mathematicians to be one of the pinnacles of mathematics! It follows from this theory that for all values of n greater than 4, the general polynomial equation of degree n cannot be solved by radicals.

Remarks on the solution of cubic equations by radicals We will sketch here Cardano's solution of cubic equations (Cardano was an Italian algebraist of the sixteenth century). Suppose our equation is $x^3 + ax^2 + bx + c = 0$. (Why can we assume that the highest coefficient is 1?) The substitution $x = y - a/3$ leads to a cubic equation $y^3 + dy + e = 0$ where there is no square term. (Work the substitution through. What are d and e in terms of a, b, and c?) Notice that if we have a root of this second equation, $y^3 + dy + e = 0$, then by subtracting $a/3$ from it, we will have a root of our original equation. At this point, we will assume that $y = u + v$, and we will separately solve for u and v. On substituting $u + v$ for y, the equation $y^3 + dy + e = 0$ is converted to $u^3 + v^3 + (d + 3uv)(u + v) + e = 0$ (check!). We have one equation in the two variables u and v, so what we will assume is that $d + 3uv = 0$, that is, we will solve for u and v with this *additional* restriction. We thus have $u^3 + v^3 = -e$ (why?) and $uv = -d/3$ (why?). Cubing this second equation, we obtain $u^3v^3 = -d^3/27$. We thus have two equations for u^3 and v^3, one describing their sum and one their product. By Exercise 12e above, we find that both u^3 and v^3 satisfy the quadratic equation $z^2 + ez - d^3/27 = 0$. Let f and g denote the roots of this equation, that is, $f = -e/2 + \sqrt{e^2/4 + d^3/27}$ and $g = -e/2 - \sqrt{e^2/4 + d^3/27}$. By symmetry, it does not matter whether we take $u^3 = f$ and $v^3 = g$ or vice versa, so we will make an arbitrary choice and take $u^3 = f$ and $v^3 = g$. By Exercise 6, f will have three cube roots, which will differ from one another by multiplication by powers of ω_3. (Here, $\omega_3 = \cos(2\pi/3) + i\sin(2\pi/3) = \dfrac{-1 + i\sqrt{3}}{2}$). Similarly, g will also have three cube roots, which will differ from one another by multiplication by powers of ω_3. We thus have three possible values of u and three possible values of v. But in addition, our u and v should satisfy $uv = -d/3$. This will not happen for every possible pairing of a cube root of f with a cube root of g. *Pick* a cube root of f and a cube root of g that multiply out to $-d/3$, and call them s and t. Then $\omega_3 s$

and $\omega_3^2 t$ will also multiply out to $-d/3$ (why?) and $\omega_3^2 s$ and $\omega_3 t$ will also multiply out to $-d/3$. Thus, our three roots of $y^3 + dy + e = 0$ are given by $y_1 = s + t$, $y_2 = \omega_3 s + \omega_3^2 t$, and $y_3 = \omega_3^2 s + \omega_3 t$, where s is a cube root of $-e/2 + \sqrt{e^2/4 + d^3/27}$ and t is a cube root of $-e/2 - \sqrt{e^2/4 + d^3/27}$, chosen so that their product equals $-d/3$.

The Field Generated by an Element

Recall the two major questions that we had posed to ourselves in Chapter 4. If K/F is a field extension and a is an element of K, we were wondering whether there is some property of a that would determine whether $F[a] = F(a)$, and we were wondering what the relation is between the element a and the degree of $F(a)$ over F. We had alluded to the minimal polynomial of a over F, but we had to delay discussing this concept until we had first studied polynomials.

Recall what it means for a to be algebraic or transcendental over F—a is algebraic over F if it satisfies some nonzero polynomial with coefficients in F, and a is transcendental over F if there is no polynomial with coefficients in F (except the zero polynomial) that is satisfied by a.

The answer to the first question is easy to describe, and in fact, we had the language with which to describe the answer in Chapter 4 itself. $F[a] = F(a)$ *if and only if* a is algebraic over F! One part of this assertion ($F[a] = F(a)$ implies that a is algebraic over F) is not difficult to prove, and we certainly could have done so in Chapter 4 itself, but the other part will take a little bit of work and will involve concepts that we developed in the chapter on polynomials. Let us prove the easier part first.

Theorem 6.1
Let K/F be a field extension, and let a be an element of K. If $F[a] = F(a)$ then a is algebraic over F.

Proof We have already seen the essential idea behind this proof in Example 4.8, where we proved that $\mathbb{R}[x^2] \neq \mathbb{R}(x^2)$.

Note that we may assume that $a \neq 0$, since if $a = 0$, then a certainly satisfies the polynomial x, which is a nonzero polynomial with coefficients in F, and is therefore definitely algebraic over F. (In fact, we have seen in Chapter 4 that more generally, *every* $a \in F$ is algebraic over F.) Now consider the element $1/a$. This is clearly an element of $F(a)$. (Is it?) Since $F(a) = F[a]$, every element in $F(a)$ can also be expressed as a polynomial expression in a with coefficients in F. Hence, $1/a$ is expressible as a polynomial expression in a with coefficients in F—say $1/a = c_0 + c_1 a + \cdots + c_k a^k$ for some $k \geq 0$ and elements $c_i \in F$. Multiplying through by a and rearranging, we obtain $c_k a^{k+1} + c_{k-1} a^k + \cdots c_0 a - 1 = 0$. But this says that a satisfies the polynomial $g(x) = c_k x^{k+1} + c_{k-1} x^k + \cdots c_0 x - 1$, which is a polynomial with coefficients in F. Moreover, $g(x) \neq 0$, since its constant coefficient is 1. Hence a is algebraic over F. □

Why did we consider the case $a = 0$ separately in the proof above?

Now how about the other direction? How do we show that if a is algebraic over F, then $F[a] = F(a)$? This is where the minimal polynomial of a over F comes in! The proper setting for the minimal polynomial is the set of *all* polynomials in $F[x]$ that are satisfied by a, and so let us consider this.

Definition 6.2
If K/F is a field extension and a is an element of K, then the set $\{f(x) \in F[x] \mid f(a) = 0\}$ is referred to as $I_{F,a}$.

(For instance, if $K = \mathbb{R}$, $F = \mathbb{Q}$, and $a = \sqrt{2}$, then $I_{\mathbb{Q},\sqrt{2}}$ contains $x^2 - 2$, $x^3 + x^2 - 2x - 2$, $x^4 - 4$, etc. Of course, $I_{\mathbb{Q},\sqrt{2}}$ will also contain the zero polynomial, since $\sqrt{2}$ rather trivially satisfies the zero polynomial!)

Observe three things immediately about $I_{F,a}$. First, what are the *constant* polynomials that $I_{F,a}$ can contain? The only constant polynomial in $I_{F,a}$ is the zero polynomial! (Why? What does it mean to substitute a for the variable x in a *constant* polynomial?) Second, what can you say about $I_{F,a}$ if a is transcendental over F? From the very definition of what it means for a to be transcendental over F, $I_{F,a}$ must contain *no* polynomial other than the zero polynomial! Third, if a is algebraic over F, how many elements should $I_{F,a}$ contain? The answer: *infinitely* many! (Why? Recall from Chapter 4 that if $f(x)$ is any nonzero polynomial satisfied by a, then for any $g(x) \in F[x]$, $g(x)f(x)$ is also satisfied by a.)

So now suppose that a is algebraic over F. Thus $I_{F,a}$ will contain infinitely many elements in it. Is there any one element in $I_{F,a}$ that somehow stands out from the rest? Certainly the zero polynomial stands out from the rest, but it is hardly an interesting polynomial! So let us rephrase the question: Is there a *nonzero* polynomial in $I_{F,a}$ that somehow stands out from the rest? Well, if such an element did exist, what possible property of this element might make it stand out? Given a set of polynomials, an obvious thing to do is to look at their degrees. Accordingly, let us look at the degrees of all the nonzero polynomials in $I_{F,a}$. Is there a single nonzero polynomial in $I_{F,a}$ of *maximum* degree? In other words, is there a positive integer n and a nonzero polynomial in $I_{F,a}$ of degree n such that every other nonzero polynomial in $I_{F,a}$ has degree less than n? If so, this polynomial will certainly stand out. But the answer is easy to see—no! (For notice that if $p(x)$ is a nonzero polynomial in $I_{F,a}$ then $xp(x)$, $x^2p(x)$, $x^3p(x)$, etc. are all in $I_{F,a}$.)

But we can look in the other direction. The degrees of the nonzero polynomials in $I_{F,a}$ will be a nonempty (why?) subset of the nonnegative integers. In fact, since the only constant polynomial in $I_{F,a}$ is the zero polynomial, the degrees of the nonzero polynomials in $I_{F,a}$ will actually be a nonempty subset of the *positive* integers. By the Well Ordering Principle, this set of degrees will have a minimal element. Let $p(x)$ be a nonzero polynomial in $I_{F,a}$ of least degree. Perhaps $p(x)$ might stand out from the rest?

But there is a problem here! $p(x)$ would certainly stand out if $p(x)$ were the *only* nonzero polynomial in $I_{F,a}$ that had the least postive degree. But what if there is *another* polynomial in $I_{F,a}$ that has the same degree as $p(x)$? Our polynomial $p(x)$ would no longer be unique. Which of these polynomials will stand out from the rest then?

As it turns out, there is a way out of this! But first, we need to study a couple of properties that are satisfied by $p(x)$, or for that matter, by any other nonzero polynomial in $I_{F,a}$ that has the least degree. Also, it would be instructive to look at an example:

Example 6.3
Let us look more carefully at the example where $K = \mathbb{R}$, $F = \mathbb{Q}$, and $a = \sqrt{2}$. What is the minimum of the degrees of the nonzero polynomials in $I_{\mathbb{Q},\sqrt{2}}$? Since $I_{\mathbb{Q},\sqrt{2}}$ already contains $x^2 - 2$, which is a polynomial of degree 2, the minimum of the degrees of the nonconstant polynomials in $I_{\mathbb{Q},\sqrt{2}}$ can only be 2 or 1. (Remember, there are no constant polynomials in $I_{\mathbb{Q},\sqrt{2}}$ other than the zero polynomial!) Let us show that it is not 1. Suppose there is a polynomial in $I_{\mathbb{Q},\sqrt{2}}$ of degree 1. It must be of the form $ax + b$ where a and b are in $\mathbb{Q}$ and $a \neq 0$. By the definition of $I_{\mathbb{Q},\sqrt{2}}$, we must get zero if we plug $\sqrt{2}$ for x in $ax + b$. Thus, we obtain $a\sqrt{2} + b = 0$, or $\sqrt{2} = -b/a$—a contradiction, as $\sqrt{2}$ is not rational. (Notice that this argument is very similar to that used in Example 5.15.3.)

Let us continue with this example. Now that we know that the minimum of the degrees of the nonzero polynomials in $I_{\mathbb{Q},\sqrt{2}}$ is 2, let us prove a couple of facts about *one* of the polynomials in $I_{\mathbb{Q},\sqrt{2}}$ that has this lowest degree, namely $x^2 - 2$. (Notice that this is not the only polynomial in $I_{\mathbb{Q},\sqrt{2}}$ that has degree 2. For instance, $2x^2 - 4$ is also in $I_{\mathbb{Q},\sqrt{2}}$, and $2x^2 - 4$ also has degree 2.) First, $x^2 - 2$ is irreducible over $\mathbb{Q}$. (Why? Recall Example 5.15.3.) Second, every polynomial in $I_{\mathbb{Q},\sqrt{2}}$ of degree 2 is of the form $k(x^2 - 2)$, where k is some rational number! (For instance, $2x^2 - 4$ is in $I_{\mathbb{Q},\sqrt{2}}$, and it is of the form $2(x^2 - 2)$.) Let us see why. Suppose $ax^2 + bx + c$ is in $I_{\mathbb{Q},\sqrt{2}}$, where a, b, and c are rational numbers and $a \neq 0$. Then $\sqrt{2}$ must satisfy this polynomial, so we find that $2a + b\sqrt{2} + c = 0$, which we rewrite as $(2a + c) + b\sqrt{2} = 0$. From the $\mathbb{Q}$-linear independence of 1 and $\sqrt{2}$, we find that $b = 0$ and $c = -2a$, that is, this polynomial is of the form $ax^2 - 2a$, or $a(x^2 - 2)$!

Now let us generalize this example. Reverting to the notation from just before this example, we will see that just as with the polynomial $x^2 - 2$ in the example, the polynomial $p(x)$ we considered before is irreducible, and that every other polynomial in $I_{F,a}$ that has the same degree as $p(x)$ is of the form $kp(x)$, where k is some (nonzero) element if F. But in the course of proving this second assertion, we will actually prove something immensely stronger!

Theorem 6.4

Let K/F be a field extension, and let a be an element of K that is algebraic over F. Let $p(x)$ be any nonzero polynomial in $I_{F,a}$ of least degree. Then

1. *$p(x)$ is irreducible, and*
2. *$p(x)$ divides* every *polynomial in $I_{F,a}$.*

(Notice how strong the second statment of the theorem is! In particular, it tells us that in our example, $x^2 - 2$ divides *every* polynomial in $I_{\mathbb{Q},\sqrt{2}}$! We could certainly have proved this while we were discussing this example above, but we have chosen to wait and prove it in the more general context of an arbitrary field extension K/F.)

Proof Let us prove the first assertion. Suppose that $p(x)$ is not irreducible. Then $p(x)$ factors as $g(x)h(x)$, where $g(x)$ and $h(x)$ are nonconstant polynomials in $F[x]$. The degree of each of $g(x)$ and $h(x)$ has to be less than that of $p(x)$. (Why?) Since $p(a) = 0$, we find (see Lemma 5.3) that $g(a)h(a) = 0$. Thus, either $g(a) = 0$ or $h(a) = 0$. Now $g(a) = 0$ tells us that $g(x)$ is in $I_{F,a}$, but this cannot be, since $deg(g(x)) < deg(p(x))$, and $p(x)$ has the least degree among all nonzero polynomials in $I_{F,a}$. Similarly, $h(a)$ cannot be zero either. It follows that $p(x)$ cannot factor nontrivially, that is, $p(x)$ must be irreducible.

Now for the second assertion. Let $f(x)$ be any element of $I_{F,a}$. By the division algorithm, there exist polynomials $g(x)$ and $r(x)$ in $F[x]$, with either $r(x) = 0$, or $0 \leq deg(r(x)) < deg(p(x))$, such that $f(x) = g(x)p(x) + r(x)$. We wish to show that $r(x)$ is zero. By Lemma 5.3, we find that $f(a) = g(a)p(a) + r(a)$. Now, $f(a)$ and $p(a)$ are both zero, since $f(x)$ and $p(x)$ are both in $I_{F,a}$. Hence $r(a)$ must also be zero. But this puts $r(x)$ in $I_{F,a}$! If $r(x)$ were not zero, then $r(x)$ would be a nonzero polynomial in $I_{F,a}$ of degree less than that of $p(x)$—a contradiction. Hence, $r(x)$ must be zero, that is, $p(x)$ must divide $f(x)$. □

We now get the following corollary almost for free!

Corollary 6.5
If $p(x)$ and $q(x)$ are any two nonzero elements of $I_{F,a}$ of least degree, then $q(x) = kp(x)$ for some nonzero $k \in F$.

Proof By the second assertion of Theorem 6.4, $p(x)$ divides $q(x)$, so $q(x) = p(x)h(x)$ for some $h(x) \in F[x]$. Since the degrees of $p(x)$ and $h(x)$ must add up to that of $q(x)$, and since the degrees of $p(x)$ and $q(x)$ are equal, we find that $deg(h(x)) = 0$, that is, $h(x)$ is a constant! Hence, $q(x)$ is just a constant multiple of $p(x)$, which is the assertion of the corollary. (Why must this constant be nonzero?) □

Here is another easy corollary:

Corollary 6.6
If $p(x)$ is any nonzero element of $I_{F,a}$ of least degree, then $I_{F,a} = \{f(x)p(x) \mid f(x) \in F[x]\}$.

Proof Theorem 6.4 showed that every polynomial in $I_{F,a}$ must be a multiple of $p(x)$, that is, $I_{F,a} \subset \{f(x)p(x) \mid f(x) \in F[x]\}$. Since $p(a) = 0$ implies $f(a)p(a) = 0$, it is clear that $f(x)p(x)$ must be in $I_{F,a}$ for any $f(x) \in F[x]$. Thus, $\{f(x)p(x) \mid f(x) \in F[x]\} \subset I_{F,a}$ as well. □

Now we are ready to pick a nonzero element of $I_{F,a}$ that will stand out from the rest. Let $p(x)$ be any nonzero polynomial in $I_{F,a}$ of least degree. Suppose $p(x) = p_n x^n + p_{n-1}x^{n-1} + \cdots + p_1 x + p_0$, with $p_n \neq 0$ (so the degree of $p(x)$ is n). Write $m_{F,a}$ for the polynomial obtained by dividing all the coefficients of $p(x)$ by p_n, that is, $m_{F,a} = x^n + (p_{n-1}/p_n)x^{n-1} + \cdots + (p_1/p_n)x + (p_0/p_n)$. Observe that $m_{F,a}$ has the same degree as $p(x)$ and that $m_{F,a}$ is *monic*, that is, its highest-degree coefficient is 1. Moreover, $m_{F,a}$ is also in $I_{F,a}$. (Why is this so?) Thus, $m_{F,a}$ has the property that it is a monic polynomial in $I_{F,a}$ of least degree. The following question immediately comes to mind: is there any other monic polynomial in $I_{F,a}$ of least degree? This is where Corollary 6.5 comes in. The answer is no! For suppose $q(x)$ is another such polynomial. Since $q(x)$ must have degree n and must be monic, $q(x) = x^n + q_{n-1}x^{n-1} + \cdots + q_1 x + q_0$ for some $q_i \in F$. By the corollary, $q(x) = c \cdot m_{F,a}$ for some $c \in F$. Comparing the coefficients of x^n in $q(x)$ and in $c \cdot m_{F,a}$, we find that $1 = c \cdot 1$, so $c = 1$. That is, $q(x) = m_{F,a}$! What we have proved is the following:

Theorem 6.7
With K, F, and a as in Theorem 6.4, $I_{F,a}$ contains exactly one monic polynomial of least degree.

Definition 6.8
If K/F is a field extension and $a \in K$ is algebraic over F, the unique monic polynomial in $I_{F,a}$ of least degree is called the *minimal polynomial of a over F*, and as in the discussion above, is denoted $m_{F,a}$.

Clearly, $m_{F,a}$ stands out among all the nonzero polynomials in $I_{F,a}$! Moreover, Theorem 6.4 guarantees that $m_{F,a}$ is irreducible and that it divides every polynomial in $I_{F,a}$.

Notice that Example 6.3 and Theorem 6.7 above together show that $m_{\mathbb{Q},\sqrt{2}}$ is $x^2 - 2$! The same techniques used in Example 6.3 will also show that the minimal polynomial of $\sqrt{p}$ over $\mathbb{Q}$, where p is any prime, is $x^2 - p$. (See also Exercise 1.)

Let us record one more useful property of $m_{F,a}$—not only is it the only monic polynomial of least degree in $I_{F,a}$, it is also the only monic *irreducible* polynomial in $I_{F,a}$! In fact, we will show something slightly stronger—any irreducible polynomial in $I_{F,a}$ must be of least postive degree.

Theorem 6.9
With K, F, and a as in Theorem 6.4, if $p(x)$ is an irreducible polynomial in $I_{F,a}$, then $p(x)$ is of least degree. If in addition, $p(x)$ is also monic, then $p(x) = m_{F,a}$.

Proof Suppose $p(x) \in I_{F,a}$ is irreducible. If $q(x)$ is any nonzero polynomial in $I_{F,a}$ of least degree (such as $m_{F,a}$), then by Theorem 6.4, $q(x)$ must divide $p(x)$, so $p(x) = q(x)h(x)$. Since $p(x)$ is irreducible, this must be a trivial factorization, so $h(x)$ must be a constant. (Why is it impossible for $q(x)$ to be a constant?) It thus follows that $p(x)$ has the same degree as $q(x)$, so by the choice of $q(x)$, $p(x)$ must also have the least degree in $I_{F,a}$. The second statement of the theorem follows from the definition of $m_{F,a}$ as *the* monic polynomial of least positive degree. □

The minimal polynomial of a over F contains the *key* to understanding the structure of the extension $F(a)/F$. For instance, there is an intimate connection between the degree of $m_{F,a}$ and the degree of the field extension $F(a)/F$—they are the same! Also, the fact that the minimal polynomial is irreducible will allow us to prove that if a is algebraic over F then $F[a]$ and $F(a)$ are equal. Thus, we would have completely answered the two questions that were left unanswered at the end of Chapter 4. Furthermore, the minimal polynomial gives us an F-basis for $F(a)$—if the degree of $m_{F,a}$ is n, then we will be able to prove that $\{1, a, a^2, \ldots, a^{n-1}\}$ is an F-basis for $F(a)$!

What is amazing is that we can obtain so much deep information about the extension $F(a)/F$ without ever leaving F. For the minimal polynomial of a is a polynomial with coefficients in F; it does *not* involve any elements outside F. Yet it gives us all these properties about an *extension* of F, that is, about a field that contains lots of elements besides the ones already in F. Clearly, this is remarkable!

So let us prove these results.

Theorem 6.10

Let K/F be a field extension, and let $a \in K$ be algebraic over F. If $deg(m_{F,a}) = n$, then

1. *$F[a] = F(a)$,*
2. *$[F(a) : F] = n$, and*
3. *The set $\{1, a, a^2, \ldots, a^{n-1}\}$ forms an F-basis for $F(a)$.*

Proof To prove 1, we need to show that $F(a) \subset F[a]$, since we already know that $F[a] \subset F(a)$. For this, it is sufficient to prove that for every polynomial $g(x) \in F[x]$ such that $g(a) \neq 0$, $1/g(a) \in F[a]$. Why? This is because every element of $F(a)$ is of the form $f(a)/g(a)$, where $f(x)$ and $g(x)$ are polynomials with coefficients in F and $g(a) \neq 0$. Hence, if $1/g(a)$ is in $F[a]$, then $1/g(a)$ would equal $h(a)$ for some polynomial $h(x) \in F[x]$, so $f(a)/g(a)$ would equal $f(a)h(a)$, which, being a product of two elements in $F[a]$, would also be in $F[a]$.

Now oberve that $m_{F,a}$ cannot divide $g(x)$, since if $g(x) = h(x)m_{F,a}$ for some $h(x) \in F[x]$, then, substituting a for x, we would find that $g(a) = h(a) \cdot 0 = 0$, while we know that $g(a) \neq 0$. By Theorem 5.17, $m_{F,a}$ and $g(x)$ must be relatively prime. Hence, there exist polynomials $h(x)$ and $f(x)$ in $F[x]$ such that $1 = h(x)m_{F,a} + f(x)g(x)$. Substituting a for x, we obtain $1 = h(0) \cdot 0 + f(a) \cdot g(a)$, that is, $f(a)g(a) = 1$. But this tells us that $1/g(a) = f(a)$. Since $f(a)$ is obtained by substituting

a for x in a polynomial with coefficients in F, $f(a)$ is an element of $F[a]$, or in other words, $1/g(a)$ is in $F[a]$!

Before we prove statements 2 and 3, note that the various powers a^i ($i = 0, 1, \ldots, n-1$) must all be distinct. For if, say, $a^i = a^j$ for $0 \leq i < j \leq n-1$), then a would satisfy $x^i - x^j$, which is a nonzero polynomial of degree *less* than that of $m_{F,a}$, a contradiction.

Now note that statement 2 will automatically follow from statement 3, since because of the distinctness of the various powers a^i, there are n elements in the set $\{1, a, a^2, \ldots, a^{n-1}\}$. Thus, it is sufficient to prove statement 3.

To prove statement 3, let us work with $F[a]$, since we *know* at this point that $F(a)$ is the same as $F[a]$. Since $F[a]$ is the set of all expressions of the form $f_0 + f_1 a + \cdots + f_k a^k$ with the $f_i \in F$ and $k \geq 0$, $F[a]$ is clearly spanned by the set consisting of all the powers of a. We need to show that we do not need *all* the powers of a to span $F[a]$, we need to show that the powers 1 through a^{n-1} will do.

For this, let $z = f_0 + f_1 a + \cdots + f_k a^k$ be an arbitrary element of $F[a]$, with the $f_i \in F$ and $k \geq 0$. Let $f(x)$ be the *polynomial* $f_0 + f_1 x + \cdots + f_k x^k$ in $F[x]$. (Recall the difference between a polynomial and a polynomial expression!) Dividing by $m_{F,a}$, we find that $f(x) = g(x) m_{F,a} + r(x)$ for some polynomials $g(x)$ and $r(x)$ in $F[x]$ with $deg(r(x)) < deg(m_{F,a})$. Suppose $r(x) = r_0 + r_1 x + \cdots r_t x^t$ for suitable $r_i \in F$ and for suitable t with $0 \leq t \leq n-1$. Substituting a for x, we obtain $z = m_{F,a}(a) g(a) + r(a) = r(a)$, so $z = r_0 + r_1 a + \cdots r_t a^t$. Thus, z is expressible as an F-linear combination of the powers 1, a, $\cdots$, a^t. As z varies through $F[a]$, the integer t will vary, but it will always be less than n, since the remainder $r(x)$ that we will get for any one z will always be of degree less than n. Thus, every element z is expressible as a linear combination of the powers 1, a, $\ldots$, a^{n-1}, so the set $\{1, a, a^2, \ldots, a^{n-1}\}$ spans $F[a]$ as an F-vector space.

Having proved that the set $\{1, a, a^2, \ldots, a^{n-1}\}$ spans $F[a]$ (which is the same as $F(a)$) as an F-space, we now need to show that this set is F-linearly independent. Assume to the contrary that some linear combination $f_0 \cdot 1 + f_1 a + \cdots + f_{n-1} a^{n-1} = 0$, where the f_i are in F, and not all the f_i are zero. But this says that a satisfies the polynomial $f(x) = f_0 + f_1 x + \cdots f_{n-1} x^{n-1}$, which is a nonzero polynomial with coefficients in F of degree less than n. Thus, $f(x)$ is a nonzero polynomial in $I_{F,a}$ of degree less than that of $m_{F,a}$—a contradiction.

Thus, the set $\{1, a, a^2, \ldots, a^{n-1}\}$ is linearly independent, and since it already spans $F(a)$ as an F-space, this set is indeed a basis for $F(a)/F$.

This proves the theorem! □

Exercises

1. Let F be a subfield of $\mathbb{C}$, and let d be an element of F that is not already a square in F (that is, there does not already exist an element y in F such that $y^2 = d$). Let $\sqrt{d}$ be any one square root of d (remember, every complex number has two square roots, and that one is the negative of the other!). Show that the minimal polynomial of $\sqrt{d}$ over F is $x^2 - d$. (Hint: Use the ideas behind Example 6.3 to show that $x^2 - d$ is irreducible over F. Now use Theorem 6.9.) It follows from Theorem 6.10 that $[F(\sqrt{d}) : F] = 2$ and that 1 and $\sqrt{d}$ form a basis for $F(\sqrt{d})$ as an F–vector space.
2. Prove that $m_{\mathbb{Q}, \sqrt[3]{2}} = x^3 - 2$. (Hint: Use Exercise 4 of Chapter 5 to show that $x^3 - 2$ is irreducible over $\mathbb{Q}$. Note that $\sqrt[3]{2}$ satisfies $x^3 - 2$. Now use Theorem 6.9.) It follows from Theorem 6.10 that $[\mathbb{Q}(\sqrt[3]{2}) : \mathbb{Q}] = 3$, and that 1, $\sqrt[3]{2}$, and $\sqrt[3]{4}$ form a basis for $\mathbb{Q}(\sqrt[3]{2})$ as a $\mathbb{Q}$–vector space.
3. In this exercise, we will prove that $m_{\mathbb{Q}, \sqrt[4]{2}} = x^4 - 2$ by showing that $x^4 - 2$ is irreducible over $\mathbb{Q}$. It will follow from Theorem 6.10 that $[\mathbb{Q}(\sqrt[4]{2}) : \mathbb{Q}] = 4$ and that 1, $\sqrt[4]{2}$, $\sqrt{2}$, and $\sqrt[4]{8}$ form a basis for $\mathbb{Q}(\sqrt[4]{2})$ as a $\mathbb{Q}$–vector space.
   (a) Use Exercise 4 of Chapter 5 to show that $x^4 - 2$ does not have a factor of degree 1. Conclude from this that $x^4 - 2$ does not have a factor of degree 3 either.
   (b) Now suppose that $x^4 - 2$ has a factor $f(x)$ of degree 2. Conclude that $\sqrt[4]{2}$ satisfies a degree 2 polynomial with coefficients in $\mathbb{Q}$.
   (c) If $\sqrt[4]{2}$ satisfies a degree 2 polynomial with coefficients in $\mathbb{Q}$, show that the linear term of this polynomial cannot be zero.
   (d) Now show that if $\sqrt[4]{2}$ satisfies a degree 2 polynomial with coefficients in $\mathbb{Q}$, then $\sqrt[4]{2} \in \mathbb{Q}(\sqrt{2})$.

(e) Finally, show that $\sqrt[4]{2} \notin \mathbb{Q}(\sqrt{2})$. (Hint: Assume that $\sqrt[4]{2} = a + b\sqrt{2}$ for some a and b in $\mathbb{Q}$. Square both sides, and use the $\mathbb{Q}$–linear independence of 1 and $\sqrt{2}$ to arrive at a contradiction.)

4. We know by now (see Example 3.11.6 of Chapter 3) that $\mathbb{Q}[\sqrt{2}, \sqrt{3}]$ is a field, with basis $\{1, \sqrt{2}, \sqrt{3}, \sqrt{6}\}$. The element $\sqrt{2} + \sqrt{3}$ is in this field, so $\mathbb{Q}(\sqrt{2} + \sqrt{3})$ is a subfield of $\mathbb{Q}[\sqrt{2}, \sqrt{3}]$. We will prove in this exercise that $\mathbb{Q}(\sqrt{2} + \sqrt{3}) = \mathbb{Q}[\sqrt{2}, \sqrt{3}]$. What this shows is that rather than first adjoining $\sqrt{2}$ and then $\sqrt{3}$ to $\mathbb{Q}$, one may obtain $\mathbb{Q}[\sqrt{2}, \sqrt{3}]$ by *directly* adjoining $\sqrt{2} + \sqrt{3}$ to $\mathbb{Q}$.

   (a) Prove using Theorem 4.2 of Chapter 4 that $[\mathbb{Q}(\sqrt{2} + \sqrt{3}) : \mathbb{Q}]$ can only be 1, 2 or 4.

   (b) Use Theorem 4.2 and Exercise 1 of Chapter 4 to show that if $[\mathbb{Q}(\sqrt{2} + \sqrt{3}) : \mathbb{Q}] = 4$ then $\mathbb{Q}(\sqrt{2} + \sqrt{3}) = \mathbb{Q}[\sqrt{2}, \sqrt{3}]$.

   (c) Use the $\mathbb{Q}$–linear independence of 1, $\sqrt{2}$, and $\sqrt{3}$ to show that $\sqrt{2} + \sqrt{3} \notin \mathbb{Q}$.

   (d) Use Exercise 1 of Chapter 4 as well as part 4c above to show that $[\mathbb{Q}(\sqrt{2} + \sqrt{3}) : \mathbb{Q}] \neq 1$.

   (e) Use Theorem 6.10 to show that if $[\mathbb{Q}(\sqrt{2} + \sqrt{3}) : \mathbb{Q}] = 2$, then $\sqrt{2} + \sqrt{3}$ must satisfy a degree 2 polynomial $x^2 + ax + b$, with a and b rational numbers.

   (f) Use the $\mathbb{Q}$–linear independence of $\{1, \sqrt{2}, \sqrt{3}, \sqrt{6}\}$ to show that a relation such as $(\sqrt{2} + \sqrt{3})^2 + a(\sqrt{2} + \sqrt{3}) + b = 0$ with a and b rational numbers is impossible.

   (g) Conclude using various parts above that $\mathbb{Q}(\sqrt{2} + \sqrt{3}) = \mathbb{Q}[\sqrt{2}, \sqrt{3}]$.

   A more general result holds: If K is *any* finite-dimensional extension of the rationals, then there exists $\alpha \in K$ such that $K = \mathbb{Q}(\alpha)$. This is known as the *Primitive Element Theorem*.

5. Suppose F, K, and L are fields with $F \subseteq K \subseteq L$. Suppose that $a \in L$ is algebraic over F. Show that a is algebraic over K and that $[K(a) : K] \leq [F(a) : F]$. (Hint: Just think very carefully about what it means for a to be algebraic over F and for a to be algebraic over K. Notice that any polynomial with coefficients in F is also a polynomial with coefficients in K.)

6. This exercise will generalize Exercise 7 of Chapter 4. In that exercise, we saw that any element $\alpha \in \mathbb{C}$ that is algebraic over $\mathbb{Q}[\sqrt{2}]$ is also algebraic over $\mathbb{Q}$. Show that more generally, if $F \subseteq K \subseteq L$ are fields and if $[K : F]$ is finite, then any element $\alpha \in L$ that is algebraic over K is also algebraic over F. (Hint: Consider the chain of fields $F \subseteq K \subseteq K(\alpha)$. What does Theorem 6.10 tell you about $[K(\alpha) : K]$?) Notice the difference between this exercise and Exercise 5 above!

7. The purpose of this exercise is to show that if K/F is a field extension and if a and b in K are both algebraic over F (with $b \neq 0$), then $a + b$, $a - b$, ab, and a/b are all algebraic over F.
   (a) Use Theorem 6.10 to show that $[F(a) : F]$ is finite.
   (b) Use Exercise 5 above to show that b is algebraic over $F(a)$.
   (c) Use Theorem 6.10 to show that $[F(a)(b) : F(a)]$ is finite.
   (d) Now use Theorem 4.2 to show that $[F(a)(b) : F]$ is finite.
   (e) Finally, use Theorem 4.12 to show that $a + b$, $a - b$, ab, and a/b are all algebraic over F.

8. Given a field extension K/F and an element $a \in K$, prove that the set $I_{F,a}$ of Definition 6.2 has the following sets of properties:
   (a) i. $0 \in I_{F,a}$.
      ii. If $f \in I_{F,a}$ then $-f \in I_{F,a}$.
      iii. If $f \in I_{F,a}$ and $g \in I_{F,a}$, then $f + g \in I_{F,a}$.
   (b) If $f \in I_{F,a}$ and if h is *any* polynomial in $F[x]$, then $hf \in I_{F,a}$.

9. The properties of $I_{F,a}$ developed in Exercise 8a above show that $(I_{F,a}, +)$ is a subgroup of $(F[x], +)$. (Of course, we have not formally considered the notion of a subgroup, but by now you should be able to formulate for yourselves the definition of a subgroup, given that you already understand the more general concepts of subrings and subspaces!) Exercise 8b, on the other hand, shows that you can multiply any element of $I_{F,a}$ by an arbitrary element of $F[x]$ and the product will still be in $I_{F,a}$. The existence of subsets in a ring that possess these two sets of properties turns out to have a deep bearing on the structure of the ring itself, and such subsets are hence studied extensively for

their own sake. They are given a name: *ideals*. The formal definition is as follows: Given a ring R, assumed to be commutative for simplicity, an *ideal* of R is a subset I of R such that

(a) $0 \in I$,

(b) $i \in I$ implies $-i \in I$,

(c) $i \in I$ and $j \in I$ implies $i + j \in I$, and

(d) for any $r \in R$, and any $i \in I$, $ri \in I$.

If R is any commutative ring, prove the following:

(a) The set $\{0\}$ is an ideal of R.

(b) R is an ideal of R.

(c) Given fixed elements $a_1, a_2, \ldots, a_n$ of R, the set $I = \{r_1a_1 + r_2a_2 + \cdots + r_na_n \mid r_1, r_2, \ldots, r_n \in R\}$ is an ideal of R. This ideal is known as the *ideal generated by* $a_1, a_2, \ldots, a_n$.

10. An ideal I of a commutative ring R is said to be *principal* if it is generated by a single element, that is, if I is of the form $\{ra \mid r \in R\}$ for some fixed $a \in R$. (By Exercise 9c above, such a set is indeed an ideal.) Notice that Corollary 6.6 above shows that ideals of $F[x]$ of the form $I_{F,a}$ are principal. Now prove that *every* ideal of $F[x]$, where F is any field, is principal. (Hint: If $I = \{0\}$, then we are done—why? Otherwise, there must be some nonzero element in I of least degree, call it q. Given an arbitrary element $f \in I$, write $f = bq + r$ (division algorithm). Use the definition of an ideal to show that r must be in I. Conclude from the choice of q and r that r must be zero. This shows that $I \subset \{rq \mid r \in F[x]\}$. Why does the reverse inclusion also hold?)

11. An ideal I of a commutative ring R is said to be *prime* if whenever $ab \in I$ for $a, b \in R$, then either $a \in I$ or $b \in I$. Given a prime integer p, show that the ideal $I = \{mp | m \in \mathbb{Z}\}$ is a prime ideal.

Notes

Remarks on Ideals (Exercise 9) Although we have focused so little on ideals in this book, they really are a very central topic in algebra, and some comments about them are in order. Historically, they gained prominence when Dedekind, while studying unique factorization in the ring of integers of an algebraic number field (see the notes on page 116), discovered that even though unique prime factorization does not always hold in such rings, the *ideals* of such rings always factored uniquely into prime ideals. (See Exercise 11 above for the definition of a prime ideal.) This turned out to be the "correct" generalization of the concept of unique prime factorization for such rings, and rings such as these where ideals factor into a product of prime ideals are now known as "Dedekind domains."

In another direction, it is easy to see that any solution of a system of polynomial equations $f_1(x_1,\ldots,x_r) = 0, \ldots, f_n(x_1,\ldots,x_r) = 0$, where the x_i are variables, will also satisfy any polynomial in the ideal of $\mathbb{C}[x_1,\ldots,x_r]$ generated by the polynomials f_i. (The ring $\mathbb{C}[x_1,\ldots,x_r]$ is just the set of polynomials in r variables with complex coefficients.) This leads to some interesting questions. One of them is the following: is every ideal of $\mathbb{C}[x_1,\ldots,x_r]$ generated by a finite set of elements? (An ideal I in a commutative ring R is said to be finitely generated if there exist elements $a_1,\ldots,a_n$ in I such that $I = \{r_1a_1 + r_2a_2 + \cdots + r_na_n \mid r_1, r_2, \ldots, r_n \in R\}$—see Exercise 9c above.) The answer is yes, and this result is known as *Hilbert's Basis Theorem*. Another question about ideals of $\mathbb{C}[x_1,\ldots,x_r]$ arises in the following context: If one starts with all solutions to a system of polynomial equations $f_1(x_1,\ldots,x_r) = 0, \ldots, f_n(x_1,\ldots,x_r) = 0$ and considers the set $\mathcal{I}$ of all polynomials in r variables that become zero at each of these solutions, then it is easy to see that $\mathcal{I}$ is actually an ideal of $\mathbb{C}[x_1,\ldots,x_r]$ that contains the ideal generated by the f_i. The question is, what is the relation between these two ideals? The answer to this is also known: For each $f \in \mathcal{I}$, some power of f must land in the ideal generated by the f_i. This result is known as *Hilbert's Nullstellensatz*.

These two examples, the factorization of ideals in the ring of integers in number fields, which is an example in algebraic number theory, and solutions of polynomial equations in r variables with complex coefficients, which is an example in algebraic geometry, were two fundamental motivators for the study of ideals in commutative rings. Ideals are now a central pillar of both algebraic number theory and algebraic geometry, and in fact actually unify the two fields into one via a mathematical object known as a *scheme*. (At this point, however, we are *hopelessly* beyond the scope of this book!)

7 CHAPTER Straightedge and Compass Constructions

Having covered an immense amount of abstract material in the previous six chapters, we are finally at the point where we can study constructibility!

While the Greeks discovered how to perform several very intricate constructions using just a straightedge and a compass, there were three very basic constructions that defied them. First, given an arbitrary angle θ, to trisect it (that is, to construct the angle $\theta/3$); second, to construct a square whose area equals that of a circle of radius one (that is, a square whose area is π); and third, to construct the side of a cube whose volume is twice that of a cube whose side is of length one (in other words, to construct the side of a cube whose volume is two). These problems were referred to (respectively) as *trisecting an arbitrary angle, squaring the circle,* and *doubling the cube.*

In the course of history, untold numbers of hours must have been spent on these problems. Clearly these problems are fascinating, and because they are easy to describe, anyone with a penchant for problem-solving can begin to attack them without any need for deep mathematical training. In fact, even though the solution to these problems is now completely known, people all around the world continue to work on these problems (perhaps unaware that the problems have been solved), and professors of mathematics constantly receive letters from the general public detailing techniques for performing one or more of the three constructions above!

These suggested techniques, of course, are all flawed, since all three constructions are impossible! What we will see in this chapter is a proof of this fact.

The proof of the impossibility of these constructions is actually very remarkable, and we should take some time out to admire several attributes of the proof. First, it is a very powerful example of the inherent worth of abstract thought. Although the problems themselves are very concrete, attempts to solve them along concrete lines failed for about two thousand years. Yet, once the more abstract methods of field theory were developed, the solution to these problems immediately fell out. Second, the proof illustrates the power and the elegance of an interdisciplinary approach to mathematics. The problems are geometric, yet the solution is purely algebraic! Some of the best mathematics of our time has been produced by combining different fields of mathematics, such as geometry and algebra, or analysis and algebra, or analysis and geometry. Third, the solution to the construction problem is *complete.* Not only does the proof give us the impossibility of the constructions, it goes further, and gives us a *verifiable* criterion for when something can be constructed. As a result, we can analyze not just the three construction problems above, but various others as well, such as the constructibility of regular n-gons, or the constructibility of angles of various measures.

The proof involves recasting the entire geometric process of straightedge and compass construction into the algebraic process of constructing field extensions of the rationals. Once this is accomplished, the criterion for when a number can be constructed can immediately be derived, and using this criterion, it is a simple matter to show that all three constructions above are impossible.

So let us look at the proof. Note first that there are precisely three elemental operations that take place during straightedge and compass constructions—either (1) we intersect two lines to determine a point, or (2) we intersect a line and a circle to determine a point, or else, (3) we intersect two circles to determine a point. Combining these three operations in various orders ultimately gives us the figure that we seek.

(For an easy example that illustrates two of these elemental operations, consider how we bisect a given line AB. With A as center, we draw a circle of large enough radius—call this circle C_1. With B as center, we similarly draw a circle of the same radius as C_1—call this circle C_2. The two circles C_1 and C_2 intersect at two points on either side of AB—call these points P_1 and P_2. Notice that the determination of P_1 and P_2 is really just operation (3) above. Next, we draw the line P_1P_2 that joins P_1 and P_2. This line intersects AB at a point M—this point is the midpoint of AB. Now notice again that this last step by which we determine M is just an instance of operation (1) above. As an exercise, you should be able to furnish for yourselves an example of a construction where all three elemental operations are used.)

What we will do is interpret each of these three operations as a process of constructing appropriate field extensions. Let us begin at the beginning. We wish to construct a particular figure using a straightedge and a compass. All we have, besides the straightedge and the compass, is a blank sheet of paper. It is very important to realize that at this stage we have no notion of length and no notion of direction at all. It is up to us to manufacture these notions.

We draw an arbitrary line on the paper. This immediately gives us a sense of direction. For instance, we could take this line to be one of two coordinate axes, for example, the x-axis. The direction perpendicular to this will then be the y-direction. We pick two distinct points P_1 and P_2 on the x-axis and construct the perpendicular bisector of P_1P_2 as described above. The perpendicular bisector becomes our y-axis, and the point of intersection of the x and y axes becomes our origin O.

We still do not have a sense of positive and negative directions on each axis, and importantly, we still do not have a notion of length. The lack of a sense of positive and negative directions is easy to fix; we simply pick any one end of the line that represents the x-axis and call it the positive x-direction. Once we have selected the positive x-direction, we pick that end of the y-axis that lies in the counterclockwise direction from the positive side of the x-axis, and call that the positive y-direction. Now how about a notion of length? Well, this is entirely up to us! We merely pick an arbitrary point P on the positive side of the x-axis, and *decree* that P is at a distance of

1 from O. Thus, P will have coordinates $(1,0)$, and the length of the line segment OP will serve as our fundamental unit of length.

(Note that during an actual construction exercise, we may not actually draw the x and y axes or pick the point P as described above. We may directly proceed to construct the figure that we want. However, it is important to realize that the process of picking a direction and a unit of length is something that happens *implicitly* when we begin our construction process. For instance, if we need to construct an equilateral triangle of side one unit, we start by drawing a line of an arbitrary length and calling that line one side of the triangle. That line immediately represents a direction on what was at first just a blank sheet of paper, and the length of the line represents one unit.)

So far, we have succeeded in constructing the coordinate axes, the origin $O = (0,0)$, and the point $P = (1,0)$. What other points on the x-axis can we construct? First of all, we can construct all points of the form $(n,0)$, with n an integer. (With O as center, draw a circle whose radius equals the length of the line segment OP, the second intersection of this circle with the x-axis will be the point $(-1,0)$. It should be clear how to get all other points on the x-axis with integer coordinates.) Repeating this procedure on the y-axis, we get all points of the form $(0,n)$, with n an arbitrary integer. Then, drawing lines parallel to the x-axis through the various points $(0,n)$ and lines parallel to the y-axis through the various points $(m,0)$, and considering all the points of intersection of these lines, we find that we can locate all points with coordinates (m,n), with m and n arbitrary integers. (It would be a good idea for you to review how to draw parallel and perpendicular lines using a straightedge and compass.)

But we can do more. Recall from geometry how to construct the length a/b $(b \neq 0)$, given that one can construct the lengths a and b (see Exercise 1). Since we can construct all integer lengths, we can use this technique to construct all lengths of the form $|n/m|$, where n and m are arbitrary integers with $m \neq 0$. (The reason for the absolute value sign above is that lengths of line segments are always nonnegative.)

We find it convenient to introduce a definition.

Definition 7.1
We say that a real number r is *constructible* if a line segment of length $|r|$ can be constructed using a straightedge and a compass.

What we have just seen is that all rational numbers are constructible.

We can now lay these (rational) lengths along the x-axis by intersecting a circle of appropriate radius centered at 0 with the x-axis. Thus, we can locate all points of the form $(q, 0)$, where q is an arbitrary rational number. Repeating this procedure along the y-axis, we find we can locate all points of the form $(0, s)$, where s is any rational number. As before, by drawing lines parallel to the two axes, we can locate all points of the form (q, s), where q and s are arbitrary rational numbers. Let us record this as a lemma:

Lemma 7.2
All points with coordinates (q, s), where q and s are arbitrary rational numbers, can be located using a straightedge and compass.

Motivated by the statement of this lemma, let us introduce some definitions.

Definition 7.3
Let F be any subfield of $\mathbb{R}$. The *plane of F* will denote the set of all points of $\mathbb{R}^2$ of the form (a, b), with a and b arbitrary elements of F (note!). We will say that *the plane of F is constructible* (or *F is constructible*) if all points from the plane of F, that is, all points of the form (a, b), with a and b arbitrary elements of F, can be located using a straightedge and compass. A *line of F* is any line whose equation can be written in the form $ax + by + c = 0$, where a, b, and c are in F. A *circle of F* is any circle whose equation can be written in the form $x^2 + y^2 + ax + by + c = 0$, where, again, a, b, and c are in F.

Understanding how one obtains the coordinates of the intersections between lines and planes of F will be crucial to showing that certain constructions are impossible. Also, notice that Lemma 7.2 effectively says that the plane of $\mathbb{Q}$ is constructible.

Can you see that if a subfield F of $\mathbb{R}$ is constructible, then $r \in F$ is also constructible?

Now let us suppose that for some subfield F of $\mathbb{R}$, the plane of F is constructible. Thus, we have a grid of points of the form (a, b), where a and b are arbitrary elements of F, each of which can be located by straightedge and compass. What do we do with these points? Well, it is quite natural that we may want to continue with our straightedge and compass constructions. Thus, we may draw lines between pairs of points in the plane of F, or circles with centers at points in the plane of F, and radii equal to the distances between various pairs of points of F, and we may intersect such lines and circles. The following lemma studies the equations of such lines and circles:

Lemma 7.4

Let P_1, P_2, and P_3 be three points in the plane of F, with P_1 and P_2 distinct, and P_2 and P_3 distinct. Let L be the line between P_1 and P_2. Also, let C be the circle with center at P_1 and radius equal to the distance between P_2 and P_3. Then L is a line of F, and C is a circle of F.

Proof Recall the definitions of "line of F" and "circle of F." Let the coordinates of P_1, P_2, and P_3 be (x_1, y_1), (x_2, y_2), and (x_3, y_3) respectively. Let us assume that $x_2 \neq x_1$. Then L has the equation $y - y_1 = \left(\frac{y_2 - y_1}{x_2 - x_1}\right)(x - x_1)$. Since x_1, x_2, y_1, and y_2 are all elements of F, it is clear that this equation is of the form $y - ax - b = 0$, where a and b are elements of F. Thus, L is a line of F. (If $x_2 = x_1$, then the line L is vertical, and has the equation $x = x_1$, so L is again a line of F.)

Now consider the circle C. Since it has a radius equal to the distance between P_2 and P_3, Pythagoras's theorem shows that the radius of C is $\sqrt{(x_3 - x_2)^2 + (y_3 - y_2)^2}$. Thus, C satisfies $(x - x_1)^2 + (y - y_1)^2 = (x_3 - x_2)^2 + (y_3 - y_2)^2$. Expanding this out, this becomes $x^2 + y^2 + ax + by + c = 0$, for suitable a, b, and c in F. Hence C is a circle of F. □

So now we have the plane of F, and we draw lines and circles using the points of the plane of F. The lemma above shows that these lines and circles are lines and circles of F, that is, their equations all have coefficients in F. What do we do with these? Recall the three elemental operations: we either intersect two of these lines, or we intersect a line and a circle, or we intersect two circles. What we are interested in is determining the precise subfield of $\mathbb{R}$ in which these coordinates lie. If we intersect a line or circle of F with another line or circle of F, will the coordinates of the point(s) of intersection be elements of F? Not necessarily. However, as the next proposition shows, if the coordinates of the point(s) of intersection do not lie in F, then, at worst, they will live in a field extension of F of the form $F(\sqrt{d})$ for some nonnegative d in F.

Proposition 7.5

Let F be a subfield of $\mathbb{R}$ that is constructible. The coordinates of the point(s) of intersection between lines and circles of F lie in a field $F(\sqrt{d})$ for some nonnegative number d in F.

Remark 7.6

The element d above depends on the particular pair of lines or circles that are being intersected. The coordinates of the points of intersection of different pairs of lines or circles could live in different field extensions ($F(\sqrt{d_1})$, $F(\sqrt{d_2})$, etc.)

Proof Suppose we are intersecting two lines of F, say L_1 and L_2. Suppose L_1 has the equation $a_1x + b_1y + c_1 = 0$, and suppose L_2 has the equation $a_2x + b_2y + c_2 = 0$, for some a_1, b_1, c_1, a_2, b_2, and c_2 in F. If the lines intersect (that is, if they are not parallel), the coordinates of the point of intersection are obtained by solving these two equations simultaneously for x and y. The coordinates are thus given by $x = \dfrac{b_1c_2 - b_2c_1}{a_1b_2 - a_2b_1}$ and $y = \dfrac{a_2c_1 - a_1c_2}{a_1b_2 - a_2b_1}$, and these coordinates clearly are elements of F.

Now suppose we are intersecting a line and a circle of F. Suppose that the line has the equation $a_1x + b_1y + c_1 = 0$, and suppose that the circle has the equation $x^2 + y^2 + a_2x + b_2y + c_2 = 0$. Then as before, the coordinates of the points of intersection (if the line and the circle intersect at all) are obtained by solving the two equations simultaneously. Observe that either a_1 or b_1 must be nonzero. (Why?) Suppose that a_1 is nonzero. Then from the equation of the line, we obtain $x = \dfrac{-b_1y - c_1}{a_1}$. Substituting this for x in the second equation,

we obtain $\left(\frac{-b_1y - c_1}{a_1}\right)^2 + y^2 + a_2\left(\frac{-b_1y - c_1}{a_1}\right) + b_2y + c_2 = 0.$
Expanding and collecting terms, we find that $ly^2 + my + n = 0$, for suitable l, m, and n in F. (Determine what l, m, and n are in terms of a_1, b_1, etc., and convince yourselves that l, m, and n are indeed in F.) This is a quadratic equation for y, whose roots are $\frac{-m \pm \sqrt{m^2 - 4ln}}{2l}$. Write d for $m^2 - 4ln$. Notice that d cannot be negative, since if it were, y would be nonreal, or in other words, our line and our circle would not intersect in $\mathbb{R}^2$. If d is a perfect square in F, then $\sqrt{d}$ will be an element of F, and it follows that the two roots will also be in F (why?). Otherwise, the two roots, being $\frac{-m \pm \sqrt{d}}{2l}$, are elements of the field $F(\sqrt{d})$. In both cases, we can think of the roots as elements of $F(\sqrt{d})$, since, after all, if d is a perfect square in F, then $F(\sqrt{d})$ is just F. Now notice that b_1, c_1, and a_1 also lie in $F(\sqrt{d})$ (why?). Hence, if y lies in $F(\sqrt{d})$, it follows from the the equation $x = \frac{-b_1y - c_1}{a_1}$ that the values of x that correspond to the two values of y also lie in $F(\sqrt{d})$. Thus, the coordinates of the points of intersection (in the case $a_1 \neq 0$) lie in the field $F(\sqrt{d})$. (Where does this process break down if $a_1 = 0$?)

The proof if $a_1 = 0$ is similar. Since b_1 cannot be zero, we can write y in terms of x from the equation of the line. We substitute for y in the equation of the circle and proceed analogously.

Now consider the case where we intersect two circles. Let the equation of the first circle be $x^2 + y^2 + a_1x + b_1y + c_1 = 0$, and let the equation of the second circle be $x^2 + y^2 + a_2x + b_2y + c_2 = 0$. Once again, we need to solve these two equations simultaneously. Subtracting the second from the first, this is equivalent to solving simultaneously the equations $(a_1 - a_2)x + (b_1 - b_2)y + (c_1 - c_2) = 0$ and $x^2 + y^2 + a_1x + b_1y + c_1 = 0$. But the first equation is just the equation of a line, so we have reduced the problem to that of intersecting a line and a circle. We have already considered this case above and have proved that the coordinates of the intersections do lie in the field $F(\sqrt{d})$ for suitable d in F. This completes the proof. □

The proof of this proposition shows that if F is a constructible field, then when intersecting lines and circles, or when intersecting two circles, we will arrive at points with coordinates in a field of the form $F(\sqrt{d})$ for some nonnegative number $d \in F$. Now let us go the other way: suppose that F is constructible, and we are given an extension field of the form $F(\sqrt{d})$, where d is some nonnegative number in F. Is $F(\sqrt{d})$ constructible? The answer is yes!

Lemma 7.7
If F is a constructible subfield of $\mathbb{R}$, then any field of the form $F(\sqrt{d})$, where d is any nonnegative number in F, is also constructible.

Proof If d is a perfect square, then $F(\sqrt{d})$ is just F, and we have nothing to prove. If not, note first that since d is constructible (why?), $\sqrt{d}$ is also constructible (see Exercise 3). By Exercise 1 in Chapter 6, every element of $F(\sqrt{d})$ can be written as $a + b\sqrt{d}$ for some a and b in F. Since b and $\sqrt{d}$ are both constructible, their product is constructible by Exercise 1b, and Exercise 1a then shows that $a + b\sqrt{d}$ must be constructible. It follows that all points of $\mathbb{R}^2$ of the form (x, y), with x and y in $F(\sqrt{d})$, can be located using a straight edge and a compass. Thus, $F(\sqrt{d})$ is constructible. □

Now we need one more observation, one that we have already made implicitly in the proof of Lemma 7.4.

Lemma 7.8
If P_1 and P_2 are points in the plane of F, then the length of the line segment P_1P_2 is an element of a field of the form $F(\sqrt{d})$ for some nonnegative $d \in F$.

Proof Pythagoras's theorem! □

We are now ready to formulate a criterion for when a real number is constructible.

Theorem 7.9
A real number α is constructible if and only if α is an element of a subfield K of $\mathbb{R}$ that has the following property: K has subfields $K_0 \subseteq K_1 \subseteq \ldots \subseteq K_t$ for some integer $t \geq 0$, where $K_0 = \mathbb{Q}$, $K_t = K$, and for $i = 1, 2, \ldots, t$, $K_i = K_{i-1}(\sqrt{d_{i-1}})$ for some nonnegative $d_{i-1} \in K_{i-1}$.

Proof Let us prove the "if" part first, that is, let us assume that α lies in a field K as described, and let us show that α must be constructible. Given the sequence of subfields $K_0 \subseteq K_1 \subseteq \ldots \subseteq K_t$ as in the statement of the theorem, we will show that each subfield in the sequence is constructible. In particular, this will show that K $(= K_t)$ must be constructible. It will then follow that α must also be constructible (why?).

When $i = 0$, K_0 is just $\mathbb{Q}$, and we have already seen (Lemma 7.2) that $\mathbb{Q}$ is constructible. Now suppose that for some $i < t$, we have already shown that K_i is constructible. Consider $K_{i+1} = K_i(\sqrt{d_i})$. If $\sqrt{d_i}$ is already in K_i, then K_{i+1} is just K_i. If not, Lemma 7.7 shows that K_{i+1} must be constructible. Thus, whenever K_i is constructible $(i < t)$, K_{i+1} is also constructible. It follows by induction that each subfield K_i $(i = 0, 1, \ldots, t)$ is constructible, as desired.

Now for the other direction: assuming that α is constructible, we must show that α must be an element of a subfield K of $\mathbb{R}$ that has the property described in the statement of the theorem. For this, we need to go back to the discussions on page 172 and study how one may construct the number α.

Recall that when we begin any construction process, we *implicitly* construct points in the plane of $\mathbb{Q}$. For example, the moment we lay out a line segment on a blank sheet of paper, that line segment represents the x-axis. If we pick a point on this line segment, perhaps by constructing the perpendicular bisector of the segment, then this point represents the origin $(0, 0)$. If we pick a second point on this line segment, then this second point represents $(1, 0)$ and the length of the segment between the two chosen points represents one unit of length. If we lay off on the perpendicular bisector a point at a distance of 1 from the origin, counterclockwise from the point $(1, 0)$, then this point represents $(0, 1)$—and so on. Now, while constructing α, we will start from points in the plane of $\mathbb{Q}$, and we may construct various lines and circles. By Lemma 7.4, these lines and circles will be in the plane of $\mathbb{Q}$. At some point in our construction, we may need to intersect some pair of lines, or a line and a circle, or a pair of circles. By Proposition 7.5, the coordinates of the intersection point will lie in a field $K_1 = \mathbb{Q}(\sqrt{d_0})$ for some nonnegative $d_0 \in \mathbb{Q}$. Note that at this point, any point, line, or circle in the plane of $\mathbb{Q}$

is also a point, line, or circle in the plane of K_1 (why?). Since our new point also has coordinates in K_1, we may think of all the points, lines, and circles that we have constructed so far as being points, lines, and circles in the plane of K_1. Now having gotten our first point of intersection, we may use this point to construct more points and more lines and circles—these points, lines, and circles will be in the plane of K_1 (why?). We may soon need to intersect another pair of lines, or a line and a circle, or two circles. By Proposition 7.5 again, the coordinates of the point of intersection will lie in the plane of a field $K_2 = K_1(\sqrt{d_1})$ for some nonnegative $d_1 \in K_1$. Once again, we may think of all the points, lines, and circles that we have constructed so far as belonging to the plane of K_2. We may now use this new point to construct more points, and more lines and circles, and these will all be in the plane of K_2. When we next intersect some pair of lines, or a line and a circle, or a pair of circles, the point of intersection will have coordinates in some field $K_3 = K_2(\sqrt{d_2})$ for some nonnegative $d_2 \in K_2$. Continuing this way, it is clear that we will arrive finally at two points P_1 and P_2, the distance between which is our number α, whose coordinates lie in some subfield K_j of $\mathbb{R}$, where K_j contains the subfields $\mathbb{Q} \subseteq K_1 \subseteq K_2 \subseteq \ldots \subseteq K_j$, and where each field in this list starting from K_1 is an extension of the previous one gotten by adjoining the square root of a nonnegative element. By Lemma 7.8, the length of P_1P_2 is an element of a field of the form $K_j(\sqrt{d_j})$ for some nonnegative $d_j \in K_j$. Since $K_j(\sqrt{d_j})$ now contains the subfields $\mathbb{Q} \subseteq K_1 \subseteq K_2 \subseteq \ldots \subseteq K_j \subseteq K_j(\sqrt{d_j})$, since each field in this new list starting from K_1 is also an extension of the previous one gotten by adjoining the square root of a nonnegative element, and since $K_j(\sqrt{d_j})$ contains the element α, we have found the desired field. □

There is an immediate corollary to this theorem that will be central to showing that certain figures are not constructible. (Note that the converse to this corollary is not true; see the notes on page 182.)

Corollary 7.10
A real number α is constructible only if it is algebraic and its minimal polynomial over $\mathbb{Q}$ is of degree a power of 2.

Proof By the previous theorem, α must be an element of a subfield K of $\mathbb{R}$, with the property described in the statement of the theorem. Each extension of the form K_i/K_{i-1} must either be of degree 1 or 2. (Why? See Exercise 1 of Chapter 6.) By repeated applications of Theorem 4.2 we find that $[K : \mathbb{Q}]$ must be a power of 2, and in particular, $K/\mathbb{Q}$ must be a finite-dimensional extension. Theorem 4.12 shows that α must be algebraic. Also, since $\mathbb{Q}(\alpha)$ is a subfield of K, Theorem 4.2 shows that $[\mathbb{Q}(\alpha) : \mathbb{Q}]$ must divide $[K : \mathbb{Q}]$, and must therefore itself be a power of 2. But $[\mathbb{Q}(\alpha) : \mathbb{Q}]$ is just the degree of the minimal polynomial of α over $\mathbb{Q}$ (Theorem 6.10.2). This gives us our result. □

We now have enough ammunition to attack our three classical constructibility problems!

Theorem 7.11
The angle 60° cannot be trisected by straightedge and compass.

Proof Assume that 60° can be trisected using a straightedge and compass. This is the same as assuming that the angle 20° can be constructed using a straightedge and compass, and by Exercise 4, this is the same as assuming that the real number $\cos 20°$ is constructible. We will use Corollary 7.10 to arrive at a contradiction.

Recall the trigonometric identity $\cos 3\theta = 4\cos^3\theta - 3\cos\theta$. Putting $\theta = 20°$, and recalling that $\cos 60° = 1/2$, we find that $\cos 20°$ satisfies $4(\cos 20°)^3 - 3(\cos 20°) = 1/2$. After clearing denominators, we find that $\cos 20°$ satisfies the polynomial $8x^3 - 6x - 1$, a polynomial that has its coefficients in $\mathbb{Q}$ (actually, in $\mathbb{Z}$ as well). By Example 5.16, this polynomial is irreducible over $\mathbb{Q}$. Since $8x^3 - 6x - 1$ is irreducible over $\mathbb{Q}$ the polynomial $x^3 - (6/8)x - (1/8)$ is also irreducible over $\mathbb{Q}$ (see Example 5.15.7). By Theorem 6.9, the minimal polynomial of $\cos 20°$ over $\mathbb{Q}$ must therefore be $x^3 - (6/8)x - (1/8)$, a cubic polynomial. But for $\cos 20°$ to be constructible, its minimal polynomial over $\mathbb{Q}$ must be of degree a power of 2 (Corollary 7.10), a contradiction! □

Theorem 7.12
Using just a straightedge and a compass, it is impossible to construct the side of a square whose area is that of a circle of radius one.

Proof We wish to construct a real number α such that $\alpha^2 = \pi$. If α were constructible, Corollary 7.10 and Theorem 6.10.2 show that $[\mathbb{Q}(\alpha) : \mathbb{Q}]$ would be finite. Since $\pi \in \mathbb{Q}(\alpha)$ (why?), Theorem 4.12 shows that π must be an algebraic number. However, although we did not prove this (a proof would be beyond the scope of this book!), we mentioned in Chapter 4 that π is *known* to be a transcendental number. Thus, α is not constructible, or in other words, it is impossible to square the circle using just a straightedge and a compass. □

Theorem 7.13
Using just a straightedge and a compass, it is impossible to construct the side of a cube whose volume is twice that of a cube whose side is of length one.

Proof We wish to show that it is impossible to construct the real number $\sqrt[3]{2}$. By Exercise 2 of Chapter 6, the minimal polynomial of $\sqrt[3]{2}$ over $\mathbb{Q}$ is $x^3 - 2$, a cubic polynomial. However, if $\sqrt[3]{2}$ were to be constructible, its minimal polynomial over $\mathbb{Q}$ must be of degree a power of two by Corollary 7.10, a contradiction. Hence, we cannot double the cube using a straightedge and a compass alone. □

Exercises

1. Let a and b be two real numbers that can be constructed using straight-edge and compass (assume $b \neq 0$).
   (a) Show that both $a + b$ and $a - b$ are constructible.
   (b) Show that ab and a/b are also constructible.
2. Let E denote the set of all real numbers that are constructible. Using Problem 1 above, show that E is a subfield of $\mathbb{R}$. This field is known as *the field of constructible numbers.*
3. If d is a positive real number that is constructible, show that $\sqrt{d}$ is also constructible.
4. Show that an angle θ can be constructed by straightedge and compass if and only if the real number $\cos\theta$ is constructible.

5. Show that if an angle θ can be constructed by straightedge and compass, then so can the angles 2θ, 3θ, ...
6. Show that a regular n-gon ($n \geq 3$) can be constructed if and only if the angle $2\pi/n$ can be constructed.
7. Show that for $k \geq 2$, the regular 2^k-gon is constructible. (Hint: Think in terms of bisecting angles!)
8. Assume that for some particular integer n ($n \geq 3$), the regular n-gon is constructible.
   (a) Show that $\cos(2\pi/n)$ must be constructible. (Hint: Combine Exercises 4 and 6 above!)
   (b) Show that for any integer $i \geq 0$, the regular $2^i n$-gon must also be constructible. (Hint: Think bisection again!)
   (c) Show that for any $k \geq 3$ that divides n, the regular k-gon must also be constructible. (Hint: Combine Exercises 5 and 6 above!)

Notes

Remarks on Corollary 7.10 It must be borne in mind that the converse to Corollary 7.10 is false: if a real number α is algebraic and its minimum polynomial over $\mathbb{Q}$ is of degree a power of 2 then α is not automatically constructible. Study closely the statement of Theorem 7.9: for α to be constructible, it is not important that $[\mathbb{Q}(\alpha) : \mathbb{Q}]$ be a power of 2 (which it certainly will be if the degree of the minimum polynomial over $\mathbb{Q}$ is a power of two); what is important is that α be an element of a field K which has a sequence of subfields as in the theorem, each of degree 2 (or 1) over the previous subfield. Now, if a field extension $K/\mathbb{Q}$ has a sequence of subfields as in the theorem, each of degree 2 (or 1), then $[K : \mathbb{Q}]$ will certainly be a power of 2 (why?). The point is that the converse of this last statement is not true—an arbitrary field extension of the rationals of degree a power of 2 need not contain such a sequence of subfields! In particular, even if $[\mathbb{Q}(\alpha) : \mathbb{Q}]$ is a power of 2, $\mathbb{Q}(\alpha)$ need not have such a sequence of subfields. Hence, we cannot conclude from the fact that $\alpha \in \mathbb{Q}(\alpha)$ and that $[\mathbb{Q}(\alpha) : \mathbb{Q}]$ is a power of 2 that α is constructible.

There is a concept of normal closure: the *normal closure of* $\mathbb{Q}(\alpha)$ *over* $\mathbb{Q}$ is the field generated over $\mathbb{Q}$ by *all* the roots of $m_{\mathbb{Q},\alpha}$. It turns out that if both $\mathbb{Q}(\alpha)$ *and* its normal closure over $\mathbb{Q}$ have degree a power of 2 over $\mathbb{Q}$, then α will be constructible.

Constructibility of n-gons We mentioned at the beginning of the chapter (see page 170) that our criterion for constructibility allows us to analyze not just the three classical problems, but also the problem of constructibility of regular n-gons for various values of n. Much as we would like to carry out this analysis, we cannot, since to do full justice to this topic, we would need to have studied roots of unity in greater detail than we have, and this would take us rather far afield.

However, the end result of the analysis is a strikingly complete condition on when a regular n-gon is constructible, and we would like to highlight the features.

One proceeds along the following lines. One shows that if a regular p-gon is constructible, where p is an odd prime, then p must be a *Fermat prime,* that is, p must be of the form $2^{2^k}+1$ for some $k \geq 0$. (Can you discover some Fermat primes less than 20?) Next, one shows that if p is an odd prime, then the regular p^2-gon is not constructible. Now one knows that if a regular n-gon is constructible, then for any $k \geq 3$ that divides n, the regular k-gon is also constructible (see Exercise 8c above), and one also knows that all regular 2^k-gons, $k \geq 2$, are constructible (see Exercise 7 above). Putting this together, one finds that if a regular n-gon ($n \geq 3$) is constructible, n must have the prime factorization $n = 2^k p_1 p_2 \cdots p_t$, where the odd primes p_i are all Fermat primes. (It is possible in this statement for n to have no odd prime factors, in which case k must be at least 2 for obvious reasons.)

But there is more, one can actually prove the converse! Thus, one shows that for any Fermat prime p, the regular p-gon is constructible. (A little Galois theory is very helpful here, though not absolutely necessary.) One also shows that if m and n are relatively prime, and if a regular m-gon and a regular n-gon are both constructible, then a regular mn-gon is also constructible. Finally, one already knows that regular 2^k-gons ($k \geq 2$) are constructible (Exercise 7), and that if a regular n-gon is constructible, then a regular $2n$-gon is also constructible (Exercise 8b). Putting all this together, we have the following result: A regular n-gon ($n \geq 3$) is constructible *if and only if* n has the prime factorization $n = 2^k p_1 p_2 \cdots p_t$, where the odd primes p_i are all Fermat primes, $t \geq 0$, $k \geq 0$ (and if $t = 0$, then $k \geq 2$).

References

Here are some other books that you might wish to consult. Keep in mind, however, that a short list such as this is necessarily incomplete, and is usually biased towards the tastes of the list-maker. In point of fact, there are *tons* of good books on all aspects of mathematics, and you might find a few afternoons spent browsing through a good library to be both pleasant and rewarding!

[1] Apostol, Tom M. 1976. *Introduction to Analytic Number Theory.* Undergraduate Texts in Mathematics. New York: Springer-Verlag.

[2] Childs, L.N. 1995. *A Concrete Introduction to Higher Algebra.* Undergraduate Texts in Mathematics. New York: Springer-Verlag.

[3] Courant, Richard and Robbins, Herbert. 1978. *What is Mathematics?* Oxford: Oxford University Press.

[4] Edwards, Harold M. 1984. *Galois Theory.* Graduate Texts in Mathematics. New York: Springer-Verlag.

[5] Hadlock, Charles Robert. 1978. *Field Theory and its Classical Problems.* Carus Mathematical Monographs. Washington D.C.: The Mathematical Association of America.

[6] Hardy, G.H. and Wright, E.M. 1979. *An Introduction to the Theory of Numbers.* Oxford: Oxford University Press.

[7] Herstein, I.N. 1975. *Topics in Algebra.* New York: John Wiley and Sons.

[8] Niven, Ivan. 1956. *Numbers: Rational and Irrational.* New Mathematical Library. Washington D.C.: The Mathematical Association of America.

[9] Niven, I. and Zuckerman, H.S. 1980. *An Introduction to the Theory of Numbers.* New York: John Wiley and Sons.

[10] Ore, Oystein. 1967. *Invitation to Number Theory.* New Mathematical Library. Washington D.C.: The Mathematical Association of America.

[11] Stillwell, J. 1994. *Elements of Algebra: Geometry, Numbers, Equations.* Undergraduate Texts in Mathematics. New York: Springer-Verlag.

[12] Stillwell, J. 1989. *Mathematics and Its History.* Undergraduate Texts in Mathematics. New York: Springer-Verlag.

Index

E

F

G

H

I

K

L

T

U

V

W

Z

Undergraduate Texts in Mathematics

(continued from page ii)